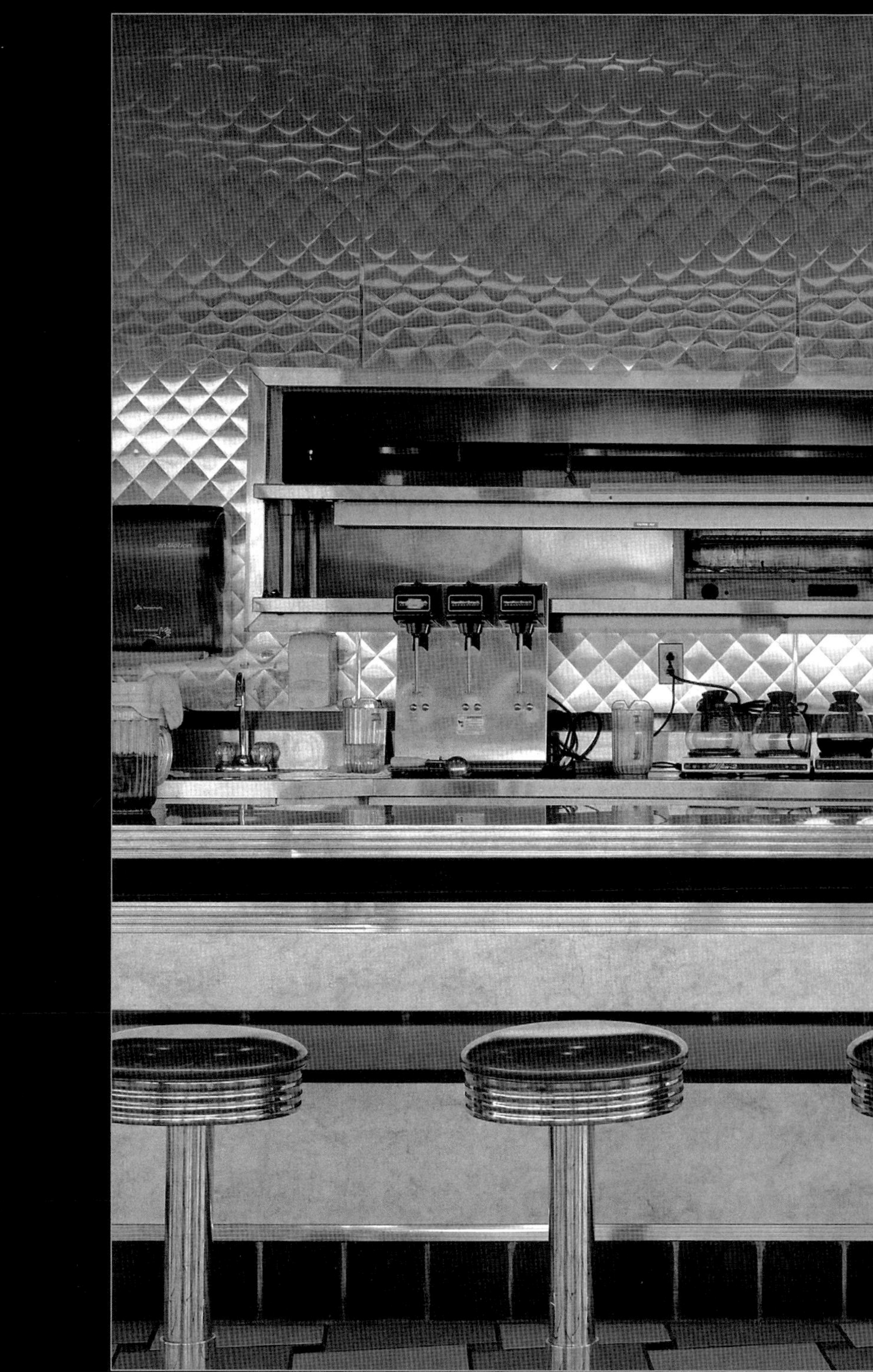

The Roadrunner Diner, Texas

Located near Amarillo, this traditional diner, resplendent with
its bar stools and stainless steel, is particularly popular with
truckers, as it offers good value, traditional American fare.

Icons of the Highway

A Celebration of Small-Town America

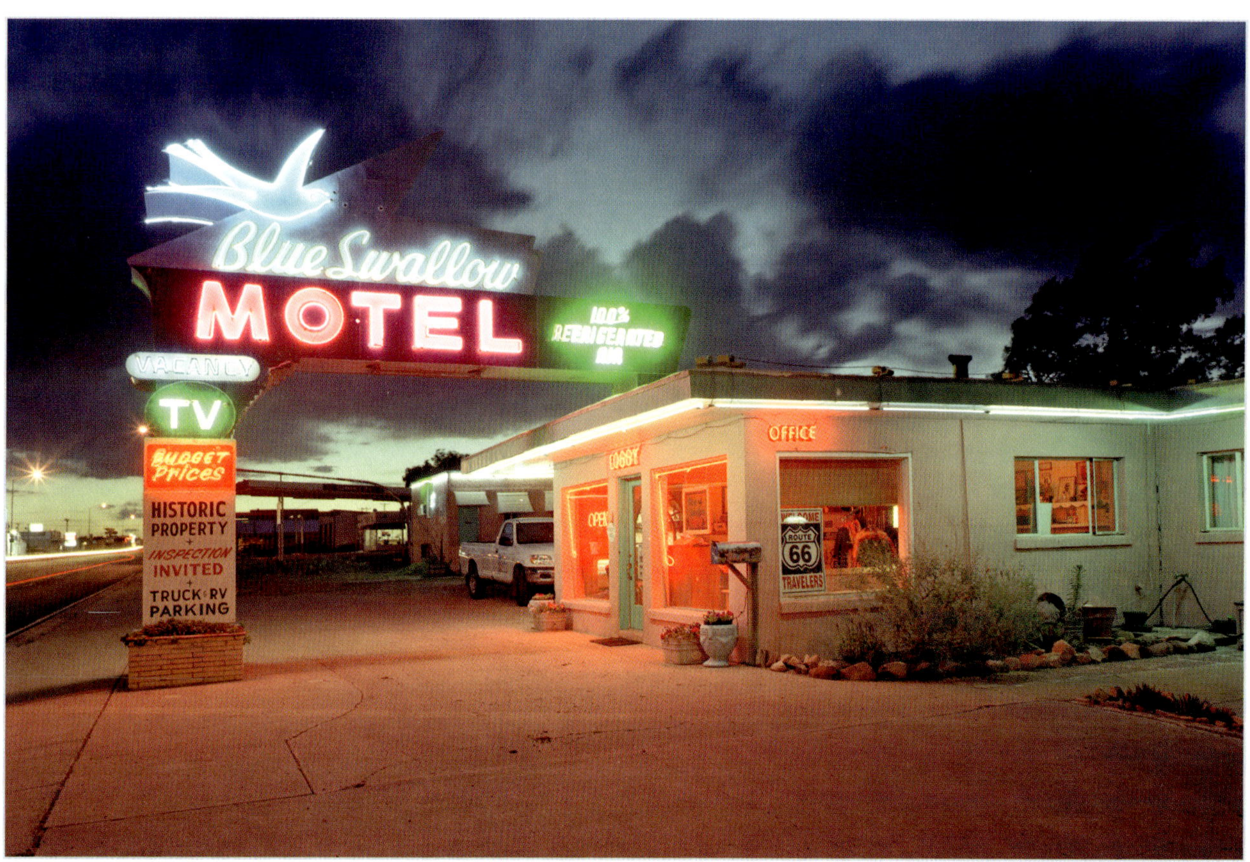

Tony & Eva Worobiec

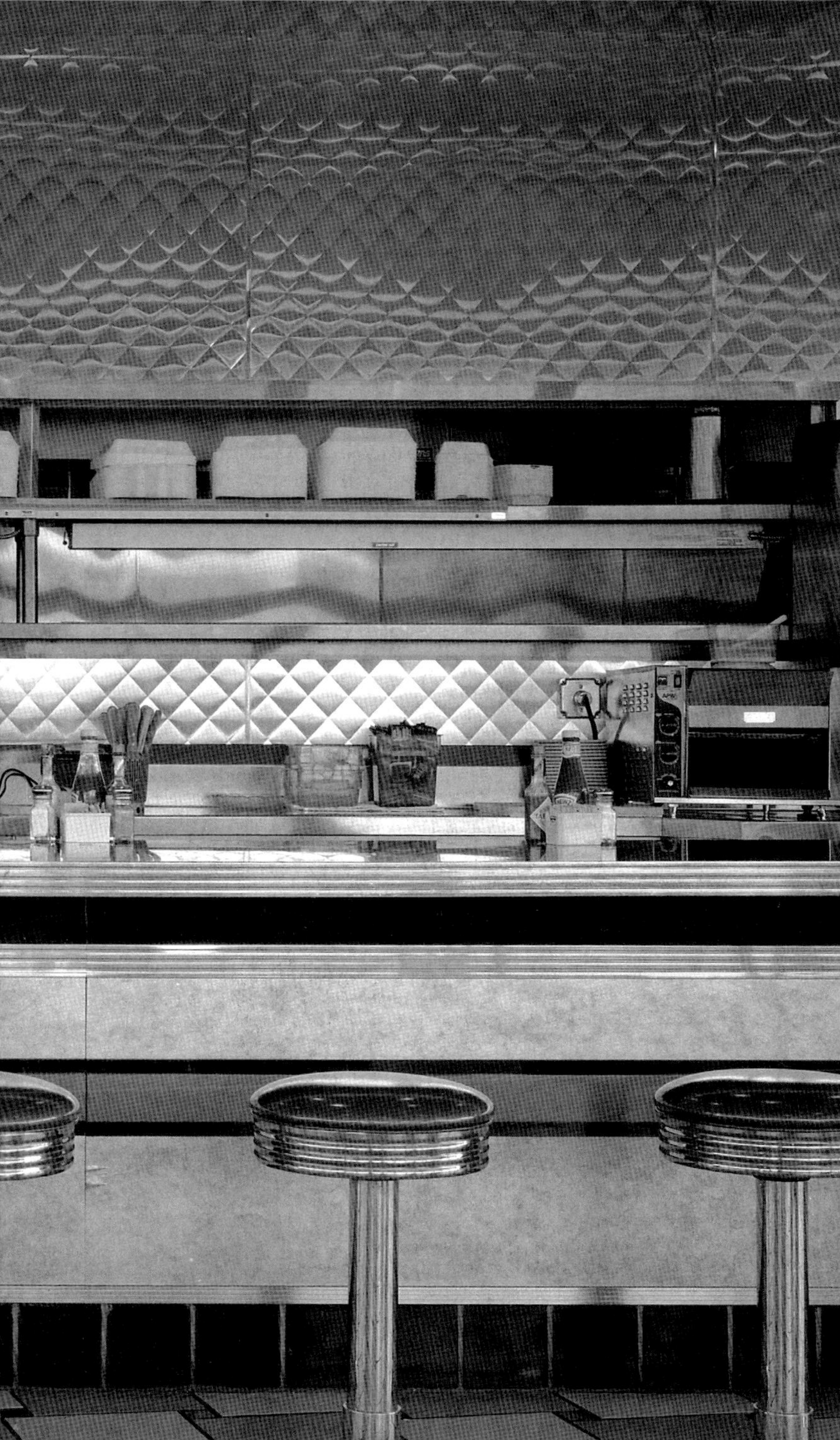

Icons of the Highway

A Celebration of Small-Town America
by Tony & Eva Worobiec

For Tricia and Tony

Icons of the Highway

A Celebration of Small-Town America
by Tony & Eva Worobiec

Published by
AAPPL Artists' and Photographers' Press Ltd.
Church Farm House, Wisley, Surrey GU23 6QL, UK
info@aappl.com   www.aappl.com

Sales and Distribution
UK and Export: Turnaround Publisher Services Ltd
orders@turnaround-uk.com
USA and Canada: Sterling Publishing Inc sales@sterlingpub.com
Australia & New Zealand: Peribo Pty Ltd michael.coffey@peribo.com.au
South Africa: Trinity Books trinity@iafrica.com

A catalogue record for this book is available from the British Library.

ISBN 9781904332787

Design (contents and cover): Stefan Nekuda office@nekuda.at

Printed in Singapore by: Imago Publishing
info@imago.co.uk

CONTENTS

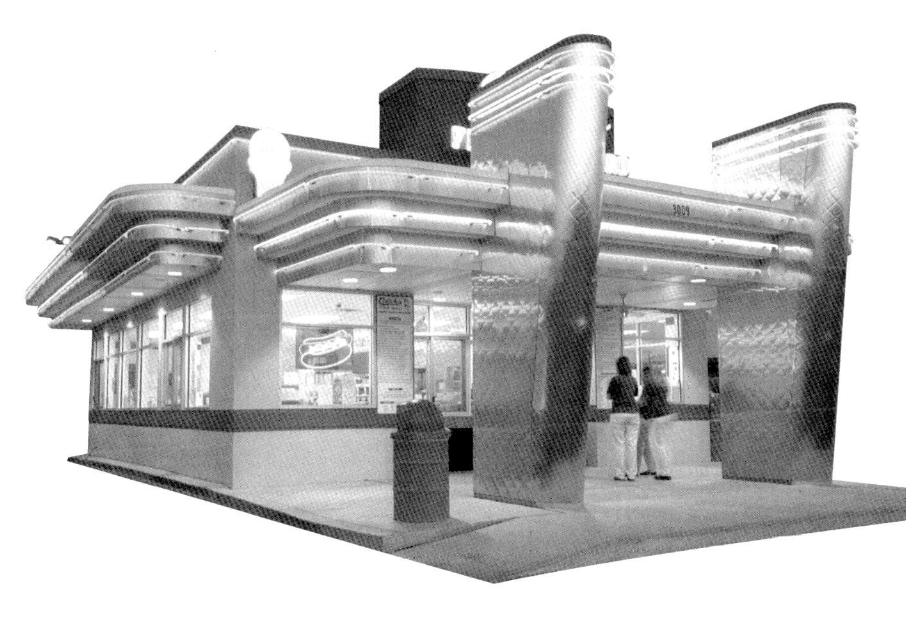

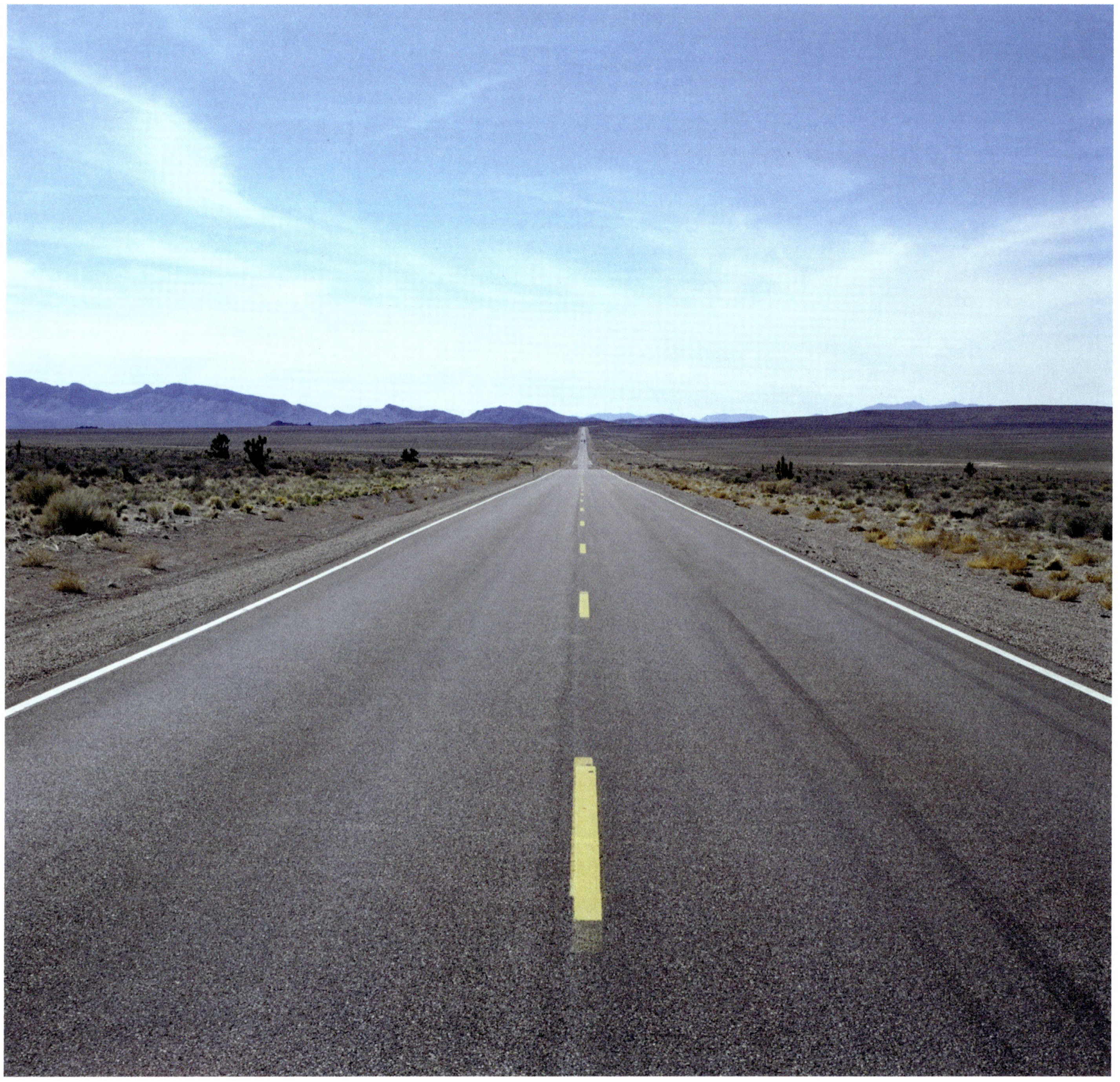

A typically dry and barren stretch of road through eastern Arizona

Introduction

Often when we think of America we remember the decades which visually set it apart – the flamboyant and oversized automobiles, the ubiquitous diners, the large city hotels wonderfully immortalized by the artist Edward Hopper, and a concept which until recent years strangely remained unique to America: the motel. In a spirit of true enterprise, each roadside establishment, be it motel, diner or cinema, would tout for custom by installing brightly colored and visually arresting neon lighting. While this tradition was later adopted by Europe and other parts of the world, the imitators could never match the audacity and enthusiasm of the American originals. The 1950s, the decade celebrated in the cult movie *American Graffiti* (1973), was perhaps the pinnacle of this extraordinary indulgence. Unlike much of the rest of the world, America had emerged from World War II in good shape. As none of its infrastructure had been damaged by bombing, it had built up an enormous industrial base in order to churn out airplanes and tanks; switching this to Buicks and Chevrolets was a relatively simple task. Most importantly, unlike Europe, America was still solvent, so its industry had more than a head start over its competitors.

In the 1950s almost every American family owned a car, as people wanted to travel – and to travel long distances. In addition, they had sufficient disposable income to eat out once in a while. This period of optimism was reflected in the upbeat designs of their vehicles, the modernistic appearance of the diners and the sleek urbanity of the motels. Love it or hate it, the Fifties style was intended to make a statement, and neon provided the gilding. Time has of course moved on; economically, much of the world has caught up. As industry becomes more global and design concepts more standardized, identifying an American car from its Japanese counterpart has

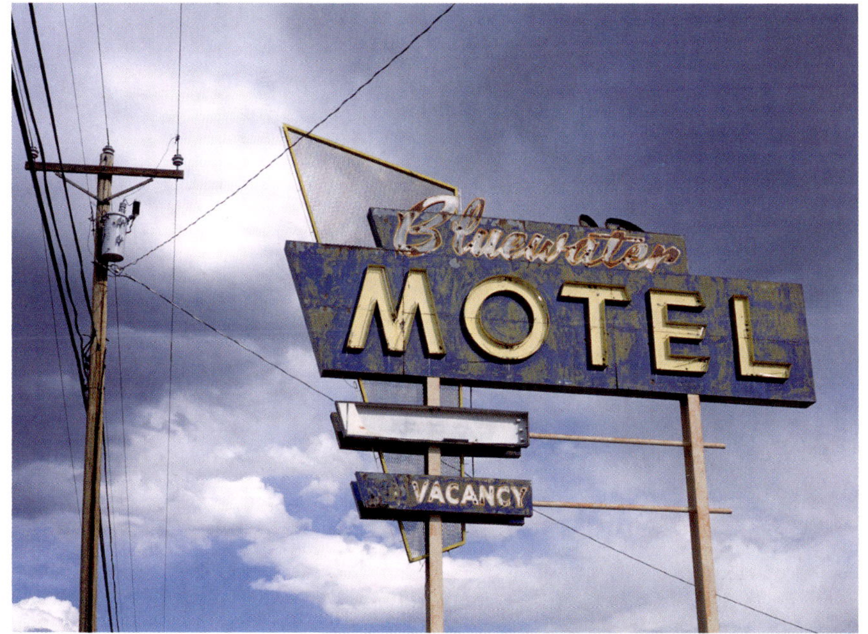

Bluewater Motel on Route 66

become ever harder. But back in the 1950s and 1960s America was unique, offering the world a culture that was steadfastly its own. Even the President and First Lady looked like A-listed movie stars. No wonder the Kennedy Administration was affectionately referred to as Camelot.

Sadly, much of what we associate with this aspect of America is now in steady decline. Changes, of course, occur everywhere, but not everywhere to the same degree. Some may regret the passing of British telephone kiosks and red pillar boxes, but most tourists revisiting Britain after a gap of twenty or thirty years will no doubt feel reassured that little has really altered in that time. By way of contrast, many American towns are being transformed entirely, and

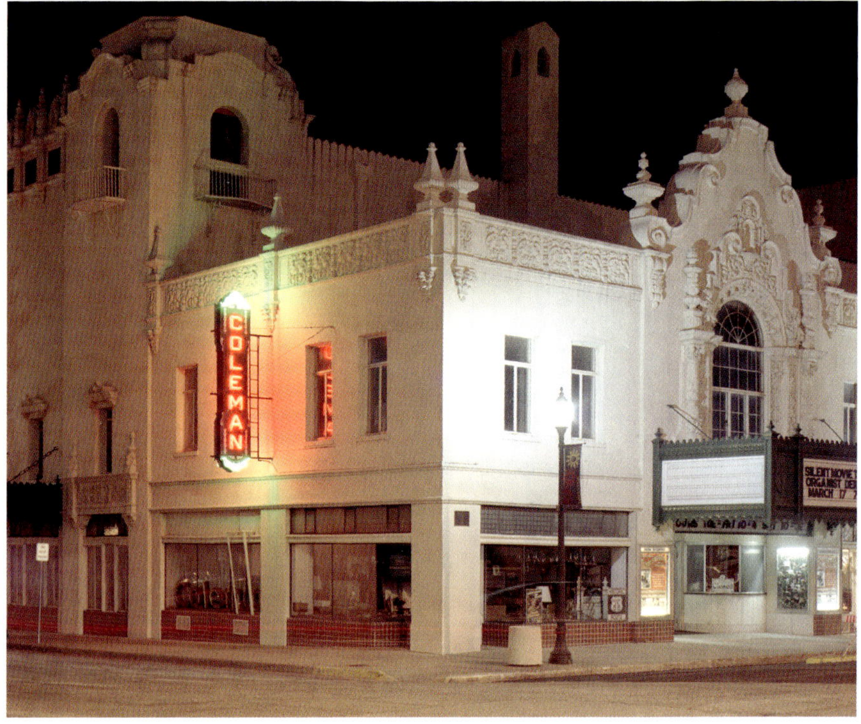

Coleman Theatre, Miami, Oklahoma

in the process quintessential American icons are at risk of being irredeemably lost.

There are of course as many hotels, motels and places to eat as there ever were, but it is the nature of these institutions that is rapidly changing. As the American Interstate network expands, more and more communities that once serviced the old highways are being bypassed. The typical traveler does not want to venture several miles off-route searching for a meal or a bed for the night. Recognizing a need, large multinational businesses have invaded the interstate intersections, offering a convenient if somewhat bland alternative to the independent diners and motels. Drive through any American town and the litany of franchises becomes bewildering; they all have their Arby's, Denny's, Wendy's and Pizza Hut. For those wishing to buy a hamburger there will always be a McDonalds or a Burger King, while Taco Bell offers something just a little more exotic. It is

almost as if the average visitor wants to be reassured that, when he walks into a restaurant, the menu on offer will be familiar. Where before part of the joy of highway eating was the constant element of surprise, now it seems every effort has been made to take the element of surprise out of it.

Remaining independent under the current economic circumstances and in the teeth of prevailing trends certainly poses difficulties. Although there are still numerous cinemas, hotels, diners and motels dotted throughout America whose owners are intent on preserving this aspect of the nation's heritage, we have been concerned, as regular visitors to the country on our various photographing trips, by how rapidly many of these independent establishments are disappearing. Motels we can recall using relatively recently have closed or been bought out by one of the franchises. Consult a guidebook printed only five years ago and you'll find that areas of downtown then famed for their neon now barely exist.

(As an aside, it's ironic that some entrepreneurs are currently building motels and diners in the 1950s style, responding to the discovery that Retro is once more in fashion. It is also hard to escape the many posters eulogizing the American heroes of the past: Bogart, JFK, the Rat Pack. Car manufacturers around the world are falling over themselves to market their latest Retro vehicle, while at least some young consumers are abandoning CDs in favor of vinyl. It is still possible to enjoy the real thing ... but for how long?)

Accordingly, for the past four years we have been methodically scouring the old American highways in search of these hauntingly beautiful places, capturing the dwindling glamour of many of the independent diners, motels, hotels, launderettes and theaters that continue to survive despite the odds. We have chosen to photograph during the twilight hour, that mystical period when day becomes night and the neon quietly appears. This work is an attempt to celebrate those iconic beacons which epitomize the American dream, in the hope that they never become just a fading memory.

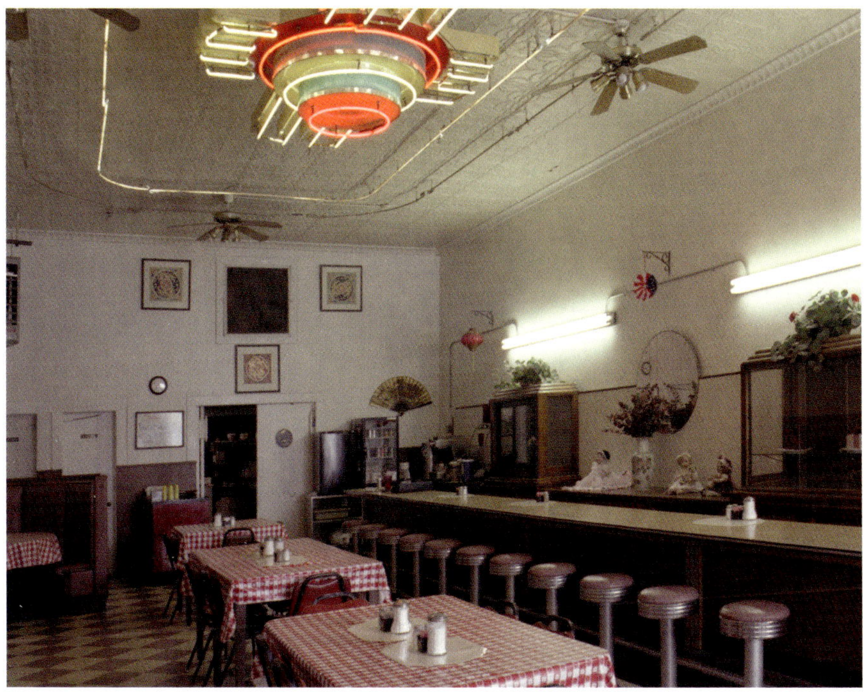

State Cafe, Las Vegas, New Mexico

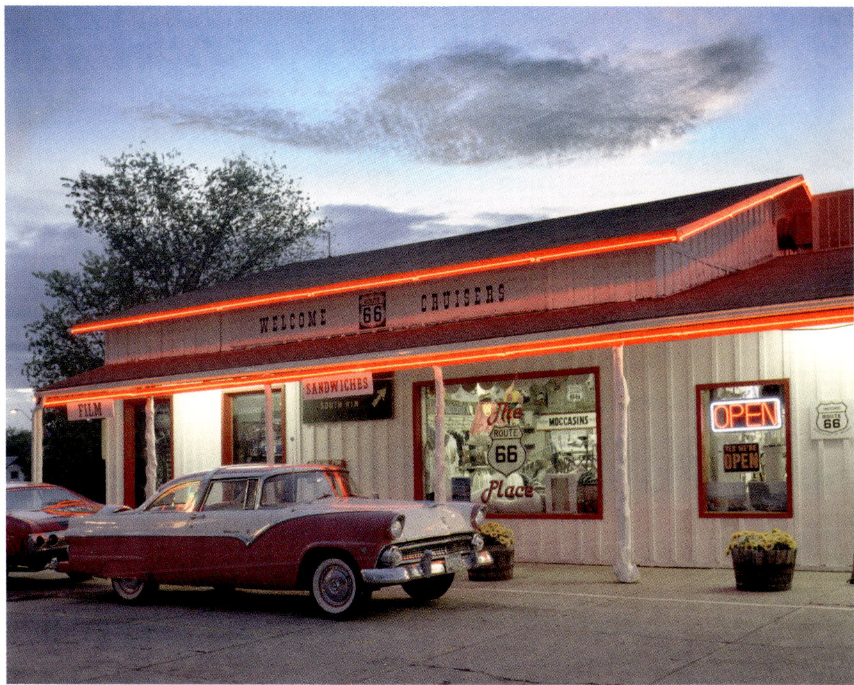

"Welcome Cruisers", Williams, Arizona

Chrome Palace: The Great American Diner

The character of any nation is immediately apparent in its cafes and restaurants. The quintessential American diner, replete with its richly padded semicircular leatherette booths, its gleaming quasi-Art Deco paneling, the mirrors, the fixed bar stools and the long, stainless-steel cafe counter, has served as a meeting point for countless generations of farmers and workers, yet far too many diners are being driven out by slick franchised establishments which insist on ushering guests to the table and presenting them with their checks almost as soon as the meal has been ordered. These are not places designed for lingering, and you find yourself hankering for a slower pace of life – for a time when you could watch the local would-be lotharios horseplay with a pretty, young waitress and her more matronly colleague, or when septuagenarian farmers in their overalls and feed caps bemoaned the diminishing returns on grain prices. Fortunately such oases of the past still exist, places where you can still sometimes find a personal mini-jukebox and almost certainly will find a menu that reflects the locality and culture. Some of the more enterprising chains have deliberately tried to recreate the spirit of the traditional diner, but it's never quite the same. The charm of the traditional diner is an informal one, one that cannot be created by conscious artifice.

For the traveler, the link between traditional eateries and the community around them is an important one. Often the business has been in the same family for generations. By their very nature, family–owned enterprises have a longstanding connection with the local area, and their staff can acquaint the passing traveler with detailed knowledge in a way the often transient workers for a large franchise can rarely match. Moreover, as in a diner that staff member who's attending to you is often the owner so the service provided is usually more personal and informal.

Despite such attractions, many of these small businesses face an uncertain future. As a single example, take the Mateo Restaurant in Santa Rosa, New Mexico, which for years has been a regular stopping point for the Greyhound Bus Company. The Mateo's proud owner, Dodie Evans, was recently told the bus is being rerouted to bypass the town. The consequences for her business are obvious. Hers is a plight shared by many all over America.

Roadside Comfort: The Family-Run Motel

When seeking accommodation for the night, the majority of travelers wish to play it safe and so choose one of the popular chains: Best Western, Day's Inn, Econolodge, and so on. The interiors of these hotels/motels are virtually identical, featuring the same decor, washroom facilities and carpets; they even smell the same. Consequently, for anyone who has been on the road for a week or more each day blurs into the last because the daily experience has been almost identical, the traveler's very own *Groundhog Day*. It's hard to ignore these places, with their large inviting driveways, free continental breakfasts, internet access and hot tubs. Finding them is never a problem: their highly familiar logos emblazon the night sky. But these signs lack the charisma, wit and glorious tackiness of those that lit up the skies of the 1950s and 1960s, when the independent motel was king.

The first motels began to appear as far back as the 1920s, but it was not until the 1950s that the concept of offering the weary traveler a home from home was fully explored. In the quintessential model, each unit was built rather like a small bungalow, with the units grouped in a horseshoe around a large courtyard. To make the task of hauling luggage easier, motels were designed so motorists could drive right up to the door – well, almost: many had a small curb a few feet from the entrance to prevent drivers from careering into the plastic chairs on either side of the door.

The best motels provided a garden and even a small swimming pool. The competition was stiff, and each vied with the next to offer the best facilities in town. TV, thick carpeting and air-conditioning came to be standard, together with bathroom facilities that would put most European hotels to shame. You only have to think of those chromium structures attached to every motel bathroom wall, supporting far more towels than even the most fastidious traveler could ever require, to fully appreciate what good value these institutions provided. Some also gave clients their own individual garage, often adjacent to their room.

Today, as customers have become more demanding, most motels offer free ice and ice bucket, a coffee machine in each room, and, as a response to the chains, a complimentary continental breakfast, albeit one usually consisting of not much more than a couple of slices of toast and a sugary donut.

In 1962, 98% of all American motels were privately owned. Since then, however, many have been swallowed up by a small number of large chains. It is still possible to find a good independent, but often their location reflects the traveling habits of previous decades. The old highways often led directly to Main Street, and these establishments were strategically placed on the arterial routes into and out of town. The arrival of the interstate, bypassing the town, has generally shifted communities toward the new route, leaving the old commercial center in danger of becoming disconnected from the mainstream of community life. The big franchises, with their huge financial clout, can not only buy up the more successful independent motels but also generate a form of commercial symbiosis whereby new corporate restaurants and cafes spring up around them, thus negating the need ever to go into town.

Many motels, particularly those in small town America, are national treasures, as they reflect a cultural heritage which is at risk of being lost. Some have changed their roles and, by dint of putting a refrigerator and a microwave in each room, now offer cheap social housing or long-term accommodation for migrant workers. Yet the independent motel is a vital part of the American culture, and it is important that it be supported and retained.

Temple of Dreams: The Independent Cinema

Cinemas have been subject to similar economic pressures. Speak to the older citizens of any community and they will tell you how the movie theater could be seen for miles because of its dazzling display of neon. But as habits have shifted, with modern audiences preferring to visit large multiplexes or rent DVDs, the character of downtown is rapidly altering. Whereas once a reasonable-size town in Texas, for example, might boast six or eight cinemas, several of them grouped around a central square, you're now lucky if you find a single such establishment surviving. Of the thirty or so original Rialtos and 55 Majestics in the state, only a couple are extant; those traditional cinemas still operating owe their current status to a committed group of local people, to the generosity of grants from arts councils, or to funding from a local authority that sees the value to the whole community of investment in the old central business area. From Artesia in New Mexico through Augusta in Kansas to Scottsbluff in Nebraska, there has been a recognition that the refurbishment of cinemas can lead to further regeneration, as more people are attracted back into the town center for an evening's entertainment. One knock-on effect is the opening or revival of cafes, shops and bars to cater for the increase in visitors.

The costs of renovation cannot be underestimated, however. The replacement of a neon or painted sign is likely to carry a price tag in the region of $25,000. An even more daunting cost for many cinemas is that of removing asbestos from their interior; in one recent example, the estimate was $300,000. As the costs for complete refurbishment of a grander establishment can rise to the millions, it is not surprising that the buildings are then often used as performing arts centers rather than just for showing movies. This is arguably no bad thing: an iconic landmark is being preserved at the same time that local people are being offered a greater choice of activities, while hopefully increased profitability is part of the equation. Ironically, the use of the premises in this new dual role mirrors their original status as venues for both film and vaudeville.

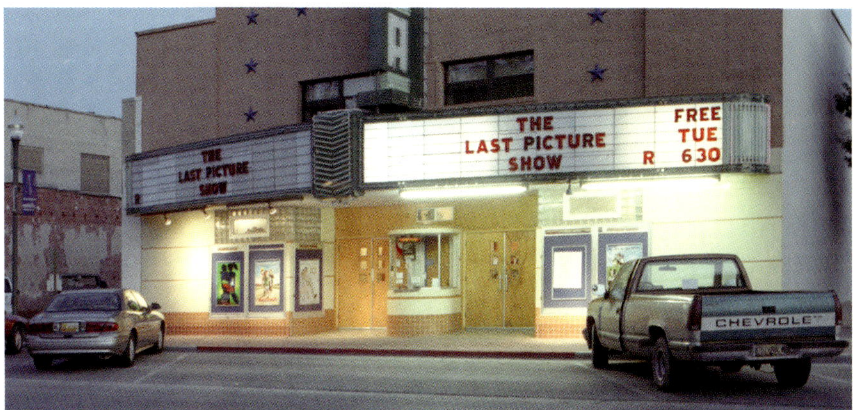

Lea Theatre, Lovington, New Mexico

Many of these picture palaces were built in the early part of the twentieth century during cycles of economic prosperity, and several still bear the name of the original benefactor. A good example is the Coleman Theatre in Miami, Oklahoma, which was built by a local magnate, George L. Coleman Sr, who had made his fortune from lead and zinc mining and wanted to provide the town with a theater as grand as anything that could be seen in any of the large cities. Other names reflect local geographical links, while the many Ritzes, Majestics, Palaces and Paramounts evoke the ambience which the original owners were no doubt hoping to create as part of the whole entertainment experience. Many of the wonderful Art Deco exteriors have been lovingly restored and preserved; the Lamar cinema in Lamar, Colorado, the Washita in Cordell, Oklahoma, and the Midwest in Scottsbluff, Nebraska, are outstanding examples. The interiors of even the smallest theaters, with their red plush seats and thick velvet curtains, retain a sense of elegance and introduce a new generation of moviegoers to the style and grace of an earlier era. When you walk into such opulent theaters as the Sooner in Norman and the Paramount in Abilene, the sheer splendor of these buildings is simply overwhelming. It is no surprise that many of these establishments have been placed on the National Register of Historic Places.

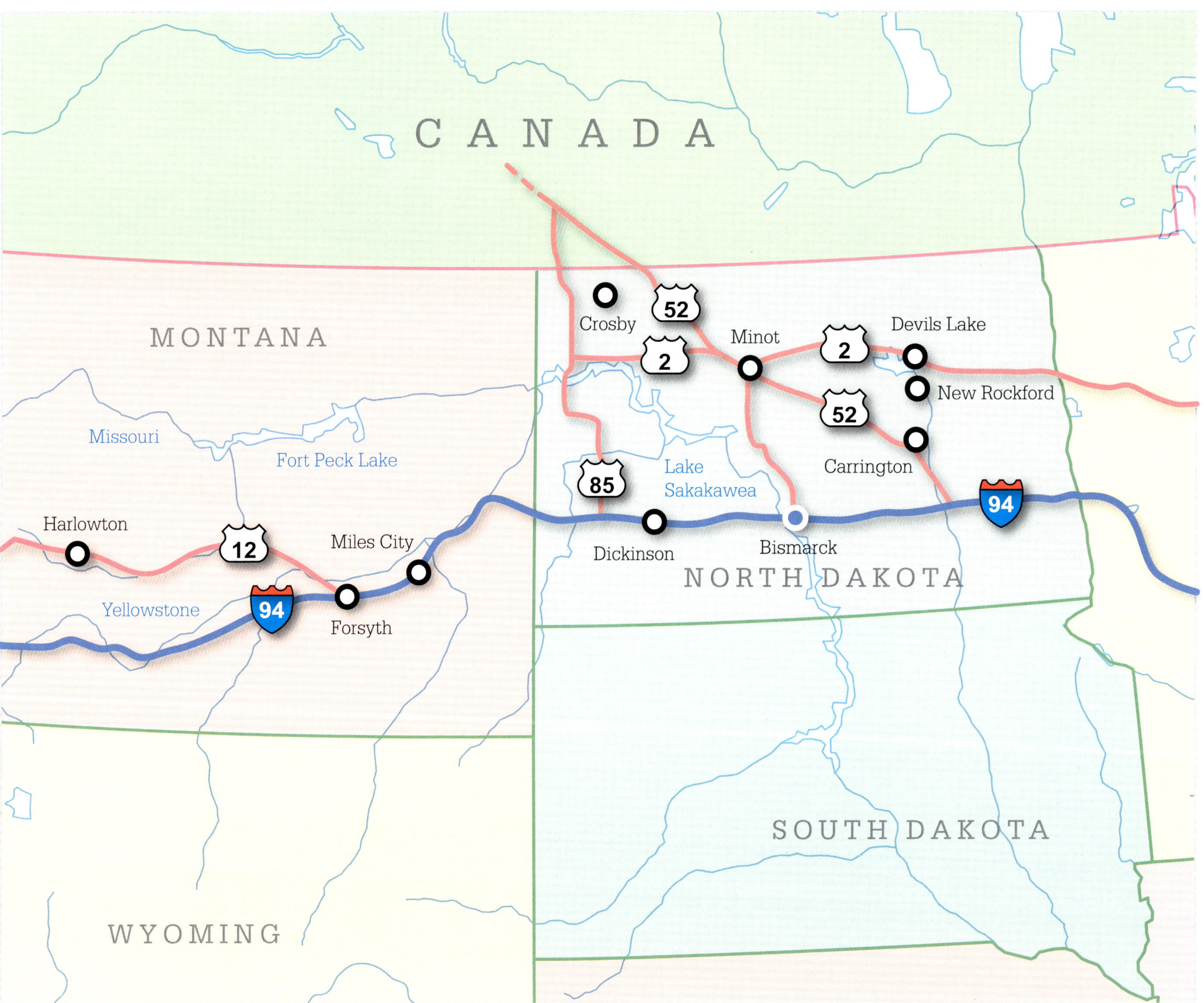

CANADA
MONTANA
Missouri
Fort Peck Lake
52
Crosby
2
Minot
2
Devils Lake
52
New Rockford
Carrington
Harlowton
12
Miles City
85
Lake Sakakawea
94
Dickinson
Bismarck
Yellowstone
94
Forsyth
NORTH DAKOTA
WYOMING
SOUTH DAKOTA

Across the High Plains

On the eastern side of the Rockies, the land descends steeply onto the windswept plains of eastern Montana. This is an area which has been appropriately named Big Sky Country, and anyone who has driven through this part of Montana will understand why; this is not a route for agoraphobics. The flat, featureless, open landscape, seemingly devoid of trees or hills, is a natural prairie where the buffalo once roamed freely. It is also a region of stunning climatic contrasts: in summer it can be baking hot while in winter it suffers Arctic blasts that come howling in straight off the Canadian plains. Residents often complain that there is no spring or fall in these parts, as winter quickly merges into summer and *vice versa*.

The terrain is a mixture of arable farmland, wide-open ranches, acres of sagebrush and strangely serrated badlands. And this area has retained some of the most quintessential mid-American towns. Especially charming is Harlowton, on the old Milwaukee line, which boasts one of the best-preserved centers anywhere in Montana; its Main Street, replete with *trompe l'oeil* facades, has been particularly well preserved.

Traveling east toward Forsyth along Highway 12, through the Musselshell Valley, the road passes by the small communities of Ryegate, Roundup, Musselshell and Melstone. Each of these towns has seen better days, and their substantial buildings evoke a prosperous past long gone. The highway weaves through numerous of the smaller farming and ranching settlements that typically sprawl across this part of Montana. West of Melstone, the arable farmland gives way to sandstone cliffs and bluffs, and, as you approach Forsyth along the banks of the Yellowstone River, sections of badland begin to emerge. This is unashamedly cowboy country. Further along the river stands the lively town of Miles City. With its handsome buildings (its Main Street has changed little since first developed near the end of the 19th century), Miles City is a natural stop-off point for many travelers

Leaving Highway 12 and venturing north, you encounter some of the bleakest landscape imaginable. While the fifty-mile gap between Miles City and Angela offers nothing but prairie, it still comes as a surprise when you get to Angela to discover that little remains of the town apart from the post office. The journey from there to Jordan offers no improvement.

In order to get up onto Highway 2, you have to do a dogleg detour around the Upper Missouri, which takes you through even more remote territory. Towns like Mosby, Grass Range and Winifred are thinly spaced out and are in every sense ranching communities. Often referred to as the Big Empty or the Big Open, this particularly desolate pocket is estimated to hold on average only one person per three square miles. Research completed by a leading American brewery established that more beer was consumed per capita in Winifred than anywhere else in America. Could these two facts be connected?

The trek across the Missouri Breaks finally leads to Highway 2, a road which runs along the roof of America and is affectionately known as the Hi-Line. The settlement of Malta very much epitomizes the type of ranchers' town that grew up alongside the Great Northern Railway. Traveling eastward, the road follows the Milk River Valley as far as Nashua, where it picks up the Missouri. The Hi-Line then passes through Glasgow, Wolf Point and, to the north, the large Fort Peck Indian Reservation.

As you cross the border into North Dakota the landscape noticeably changes. The highways taper to infinity past enormous

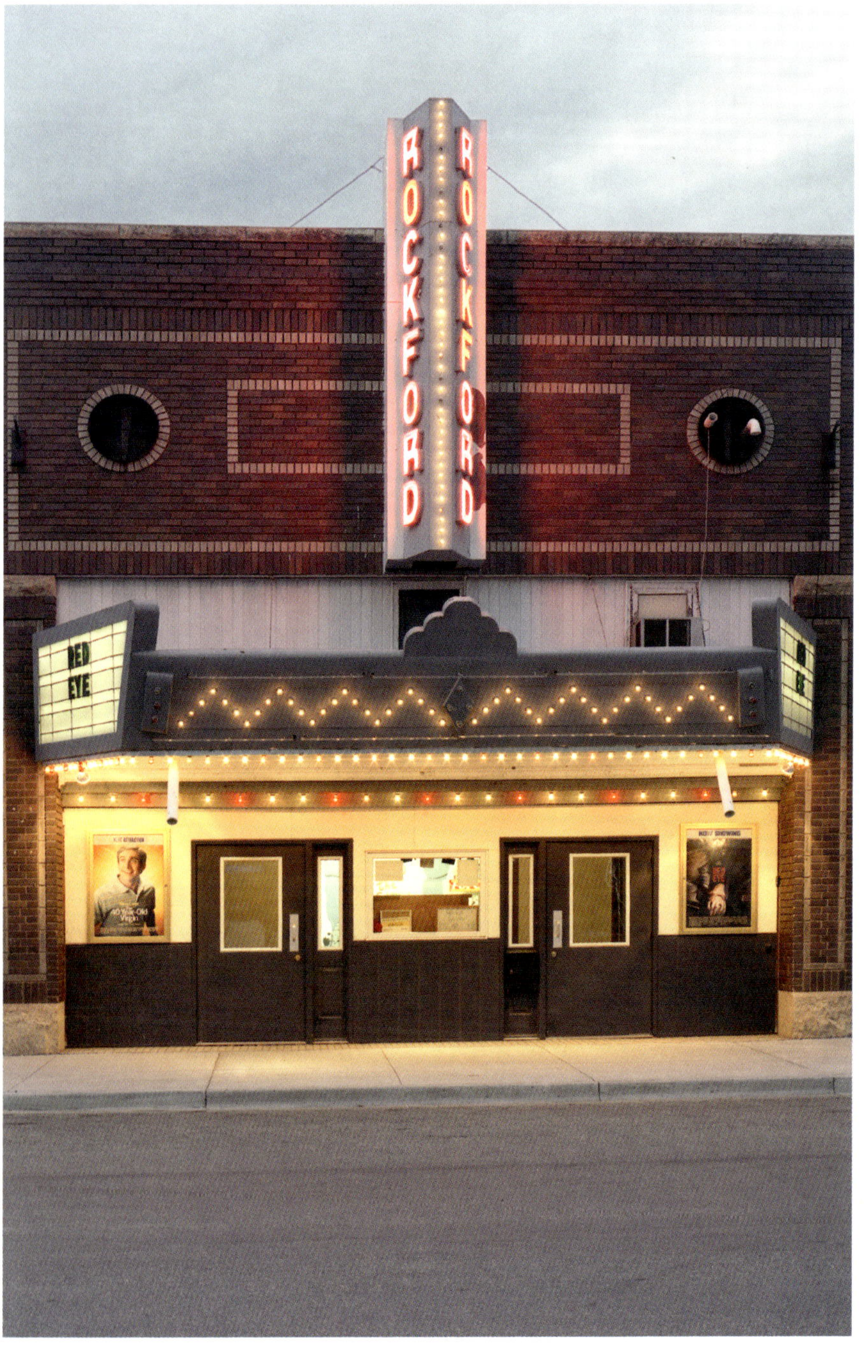

The Rockford Theatre, New Rockford, North Dakota

fields of cereal fanned into waves by the wind that blows over the prairie. Occasionally the road dips to bisect one of the countless small lakes that seem to litter this otherwise featureless part of America. While it might be thinly populated, North Dakota enjoys a rich wildlife and is a paradise for ornithologists; it is not unusual to see a group of herons searching for fish by the water's edge. Williston, once described as "a one-movie-theater town", typifies this northeastern corner of the Dakotas. It prospered on oil pumping and wheat growing, but has learned to diversify.

As you travel further east, past tiny communities like Ross and Stanley, small hillocks appear, residual remnants of the last Great Ice Age. The only town of any substance is Minot, another settlement that expanded rapidly with the development of the Great Northern Railway. Continuing along Highway 2, the fields of cereals and hay seem even vaster, interrupted only by the occasional patch of sunflowers. Despite the English origins of so many of the place-names, the population is decidedly Scandinavian, and in the cafes of the area it's not uncommon to be offered smorgasbord. At Rugby, you have hit the geographical center of North America.

Sixty miles east of Rugby lies the charming town of Devils Lake. Built north of an area of wetlands, it is a favorite haunt for hunters and fisherman. South and east of Devils Lake is the equally appealing little North Dakota town of Rockford, which, through the enthusiastic support of a small, tight-knit community, has kept its wonderful movie theater alive.

For those accustomed to urban snarl-ups, driving across the High Plains is a liberating experience. You can travel great distances without ever seeing another vehicle, while the thinly distributed towns retain many features that recall the decades of the early and middle twentieth century.

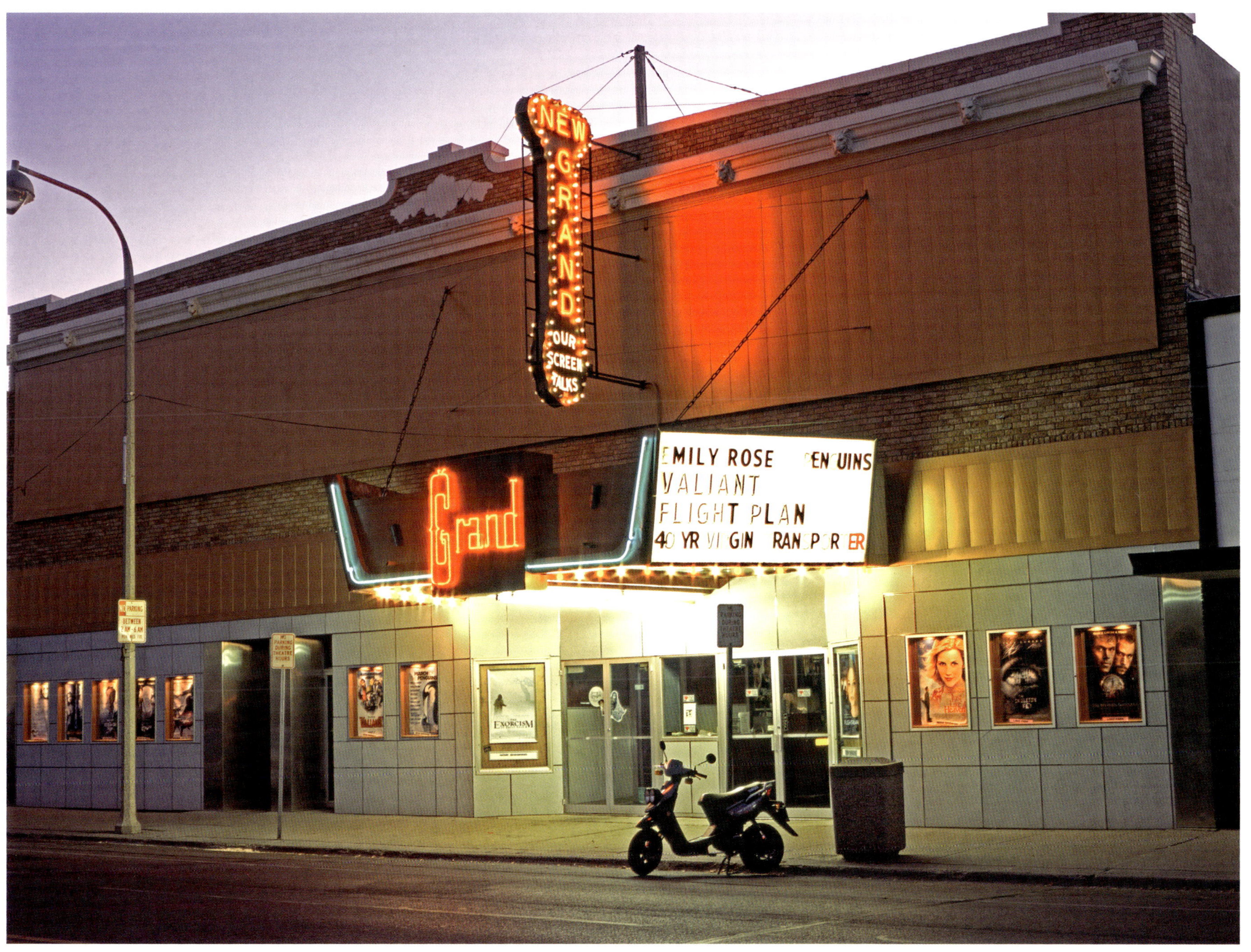

The New Grand Theatre, Williston, North Dakota

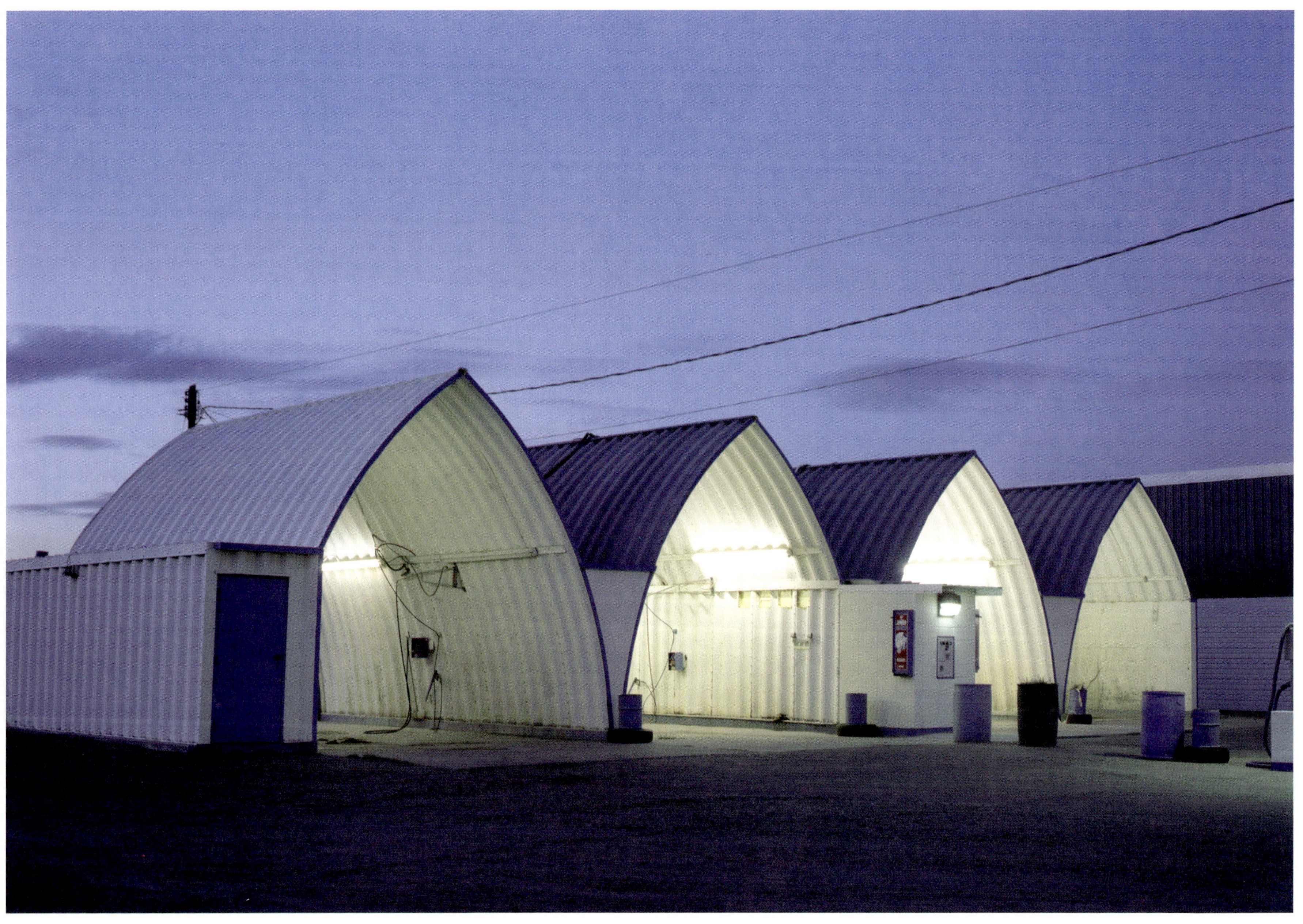

Car-wash, Forsyth, Montana

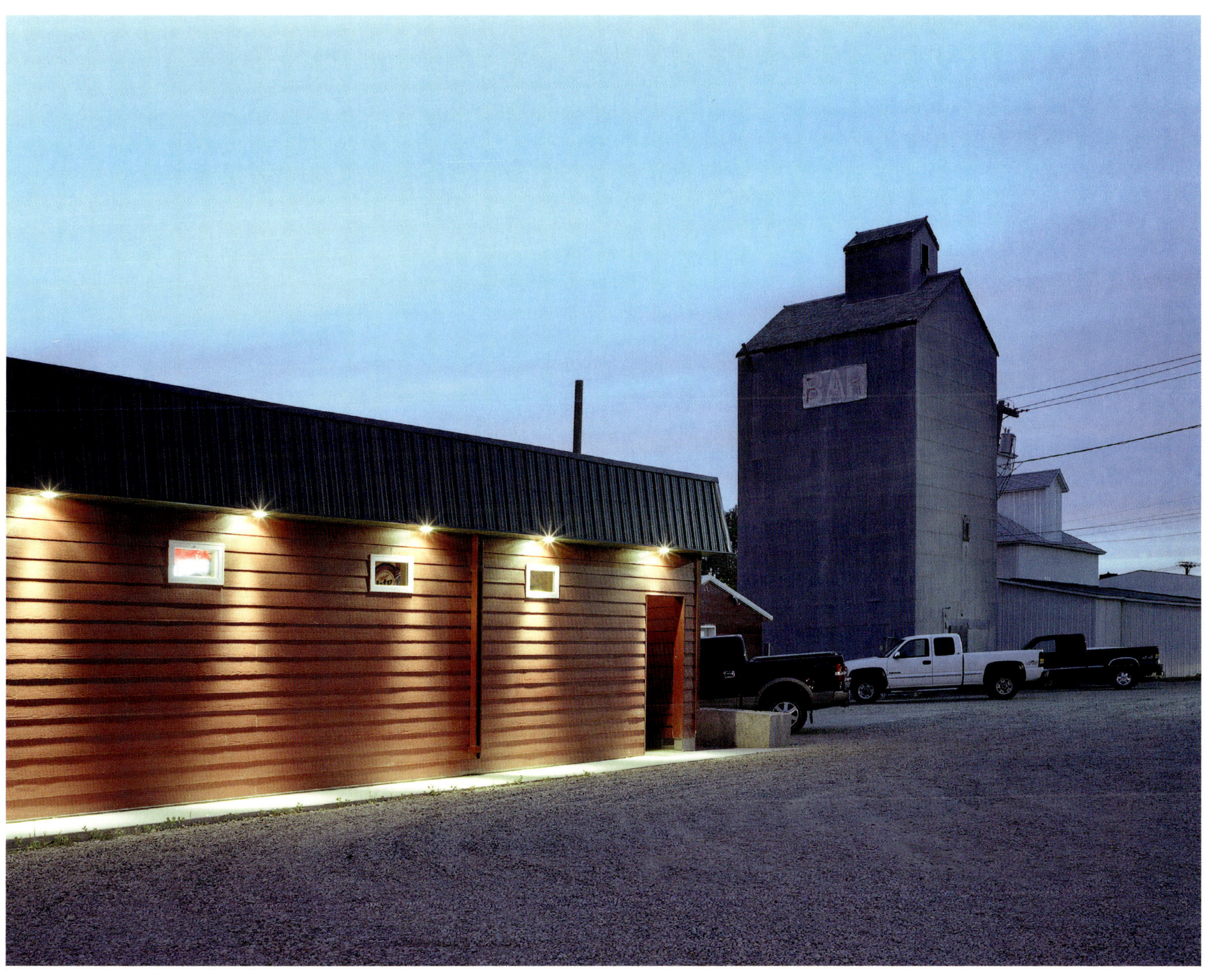

Bar, Devil's Lake, North Dakota

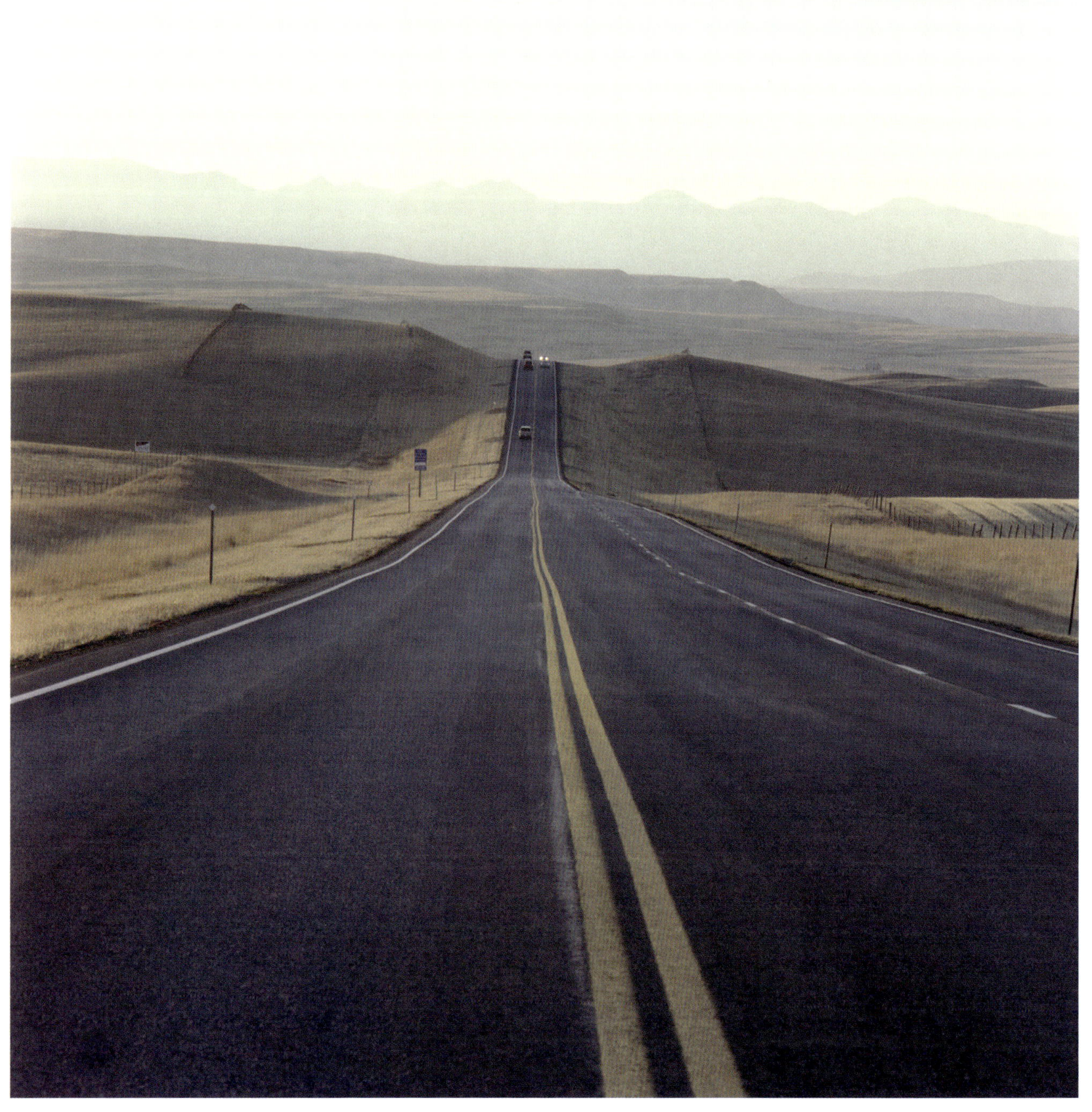

Hwy 191, north of Harlowton, Montana

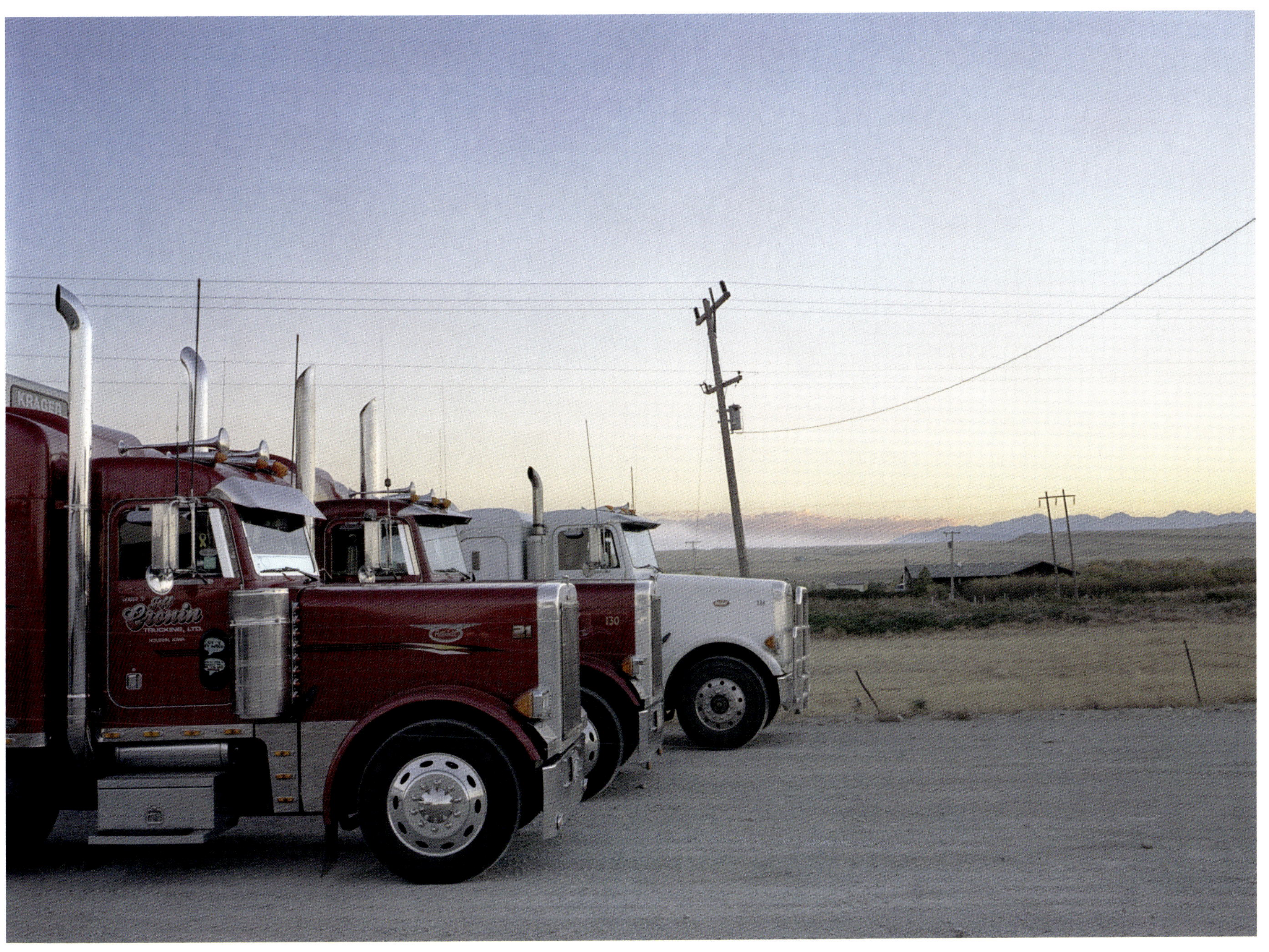

Trucks parked up for the night, opposite the Corral Motel, Harlowton, Montana

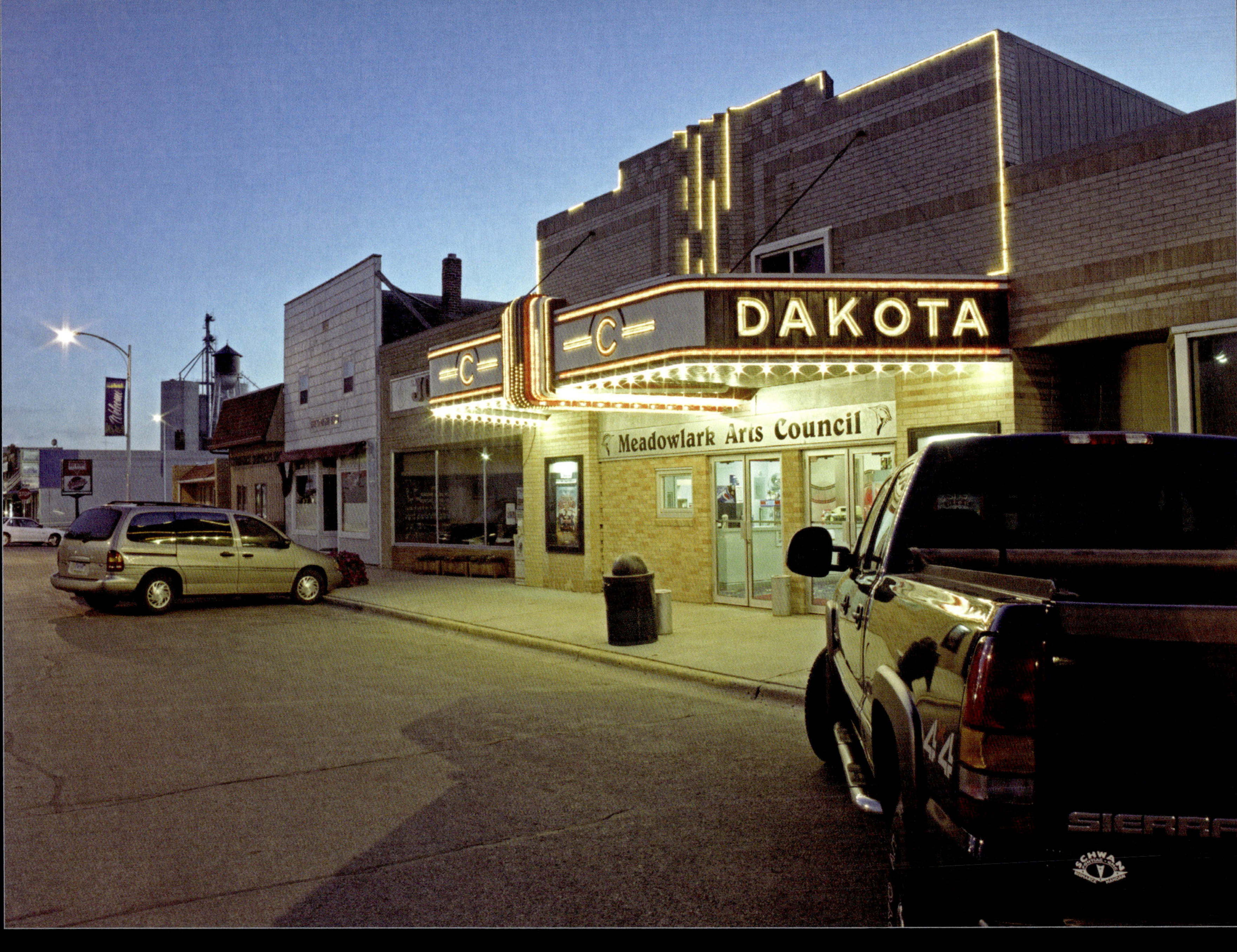

The Dakota Theatre, Crosby, North Dakota

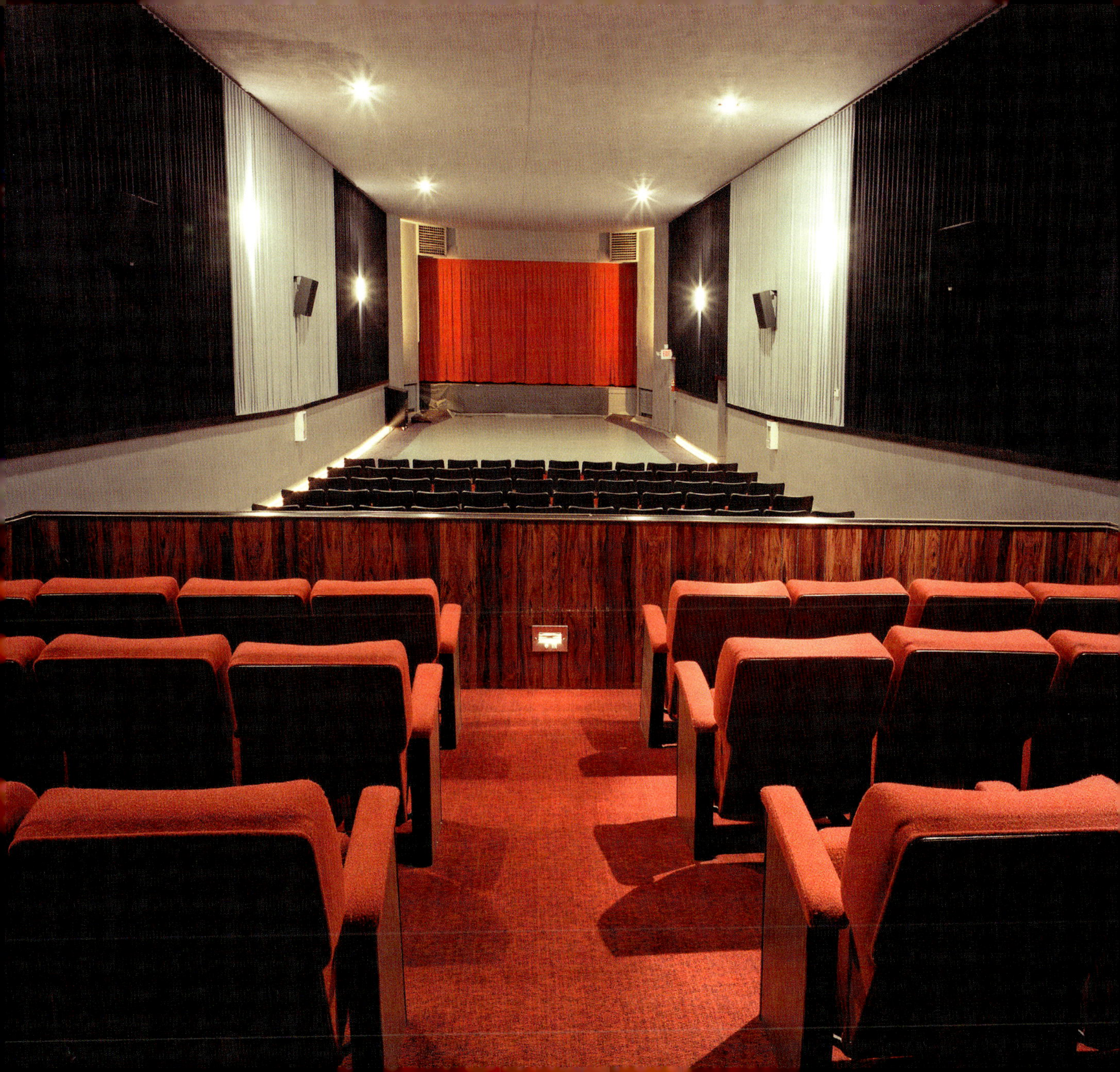

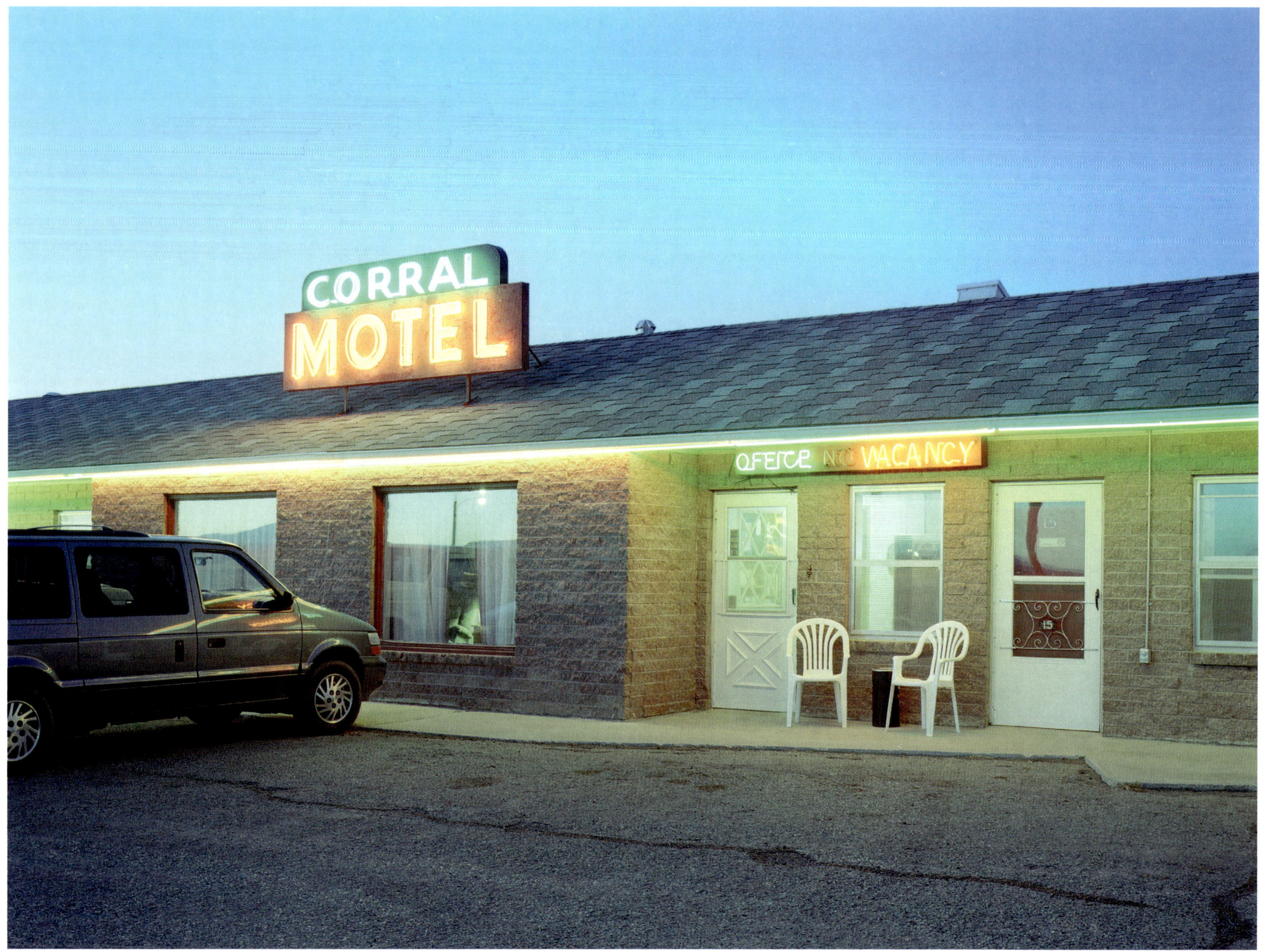

The Corral Motel, Harlowton, Montana

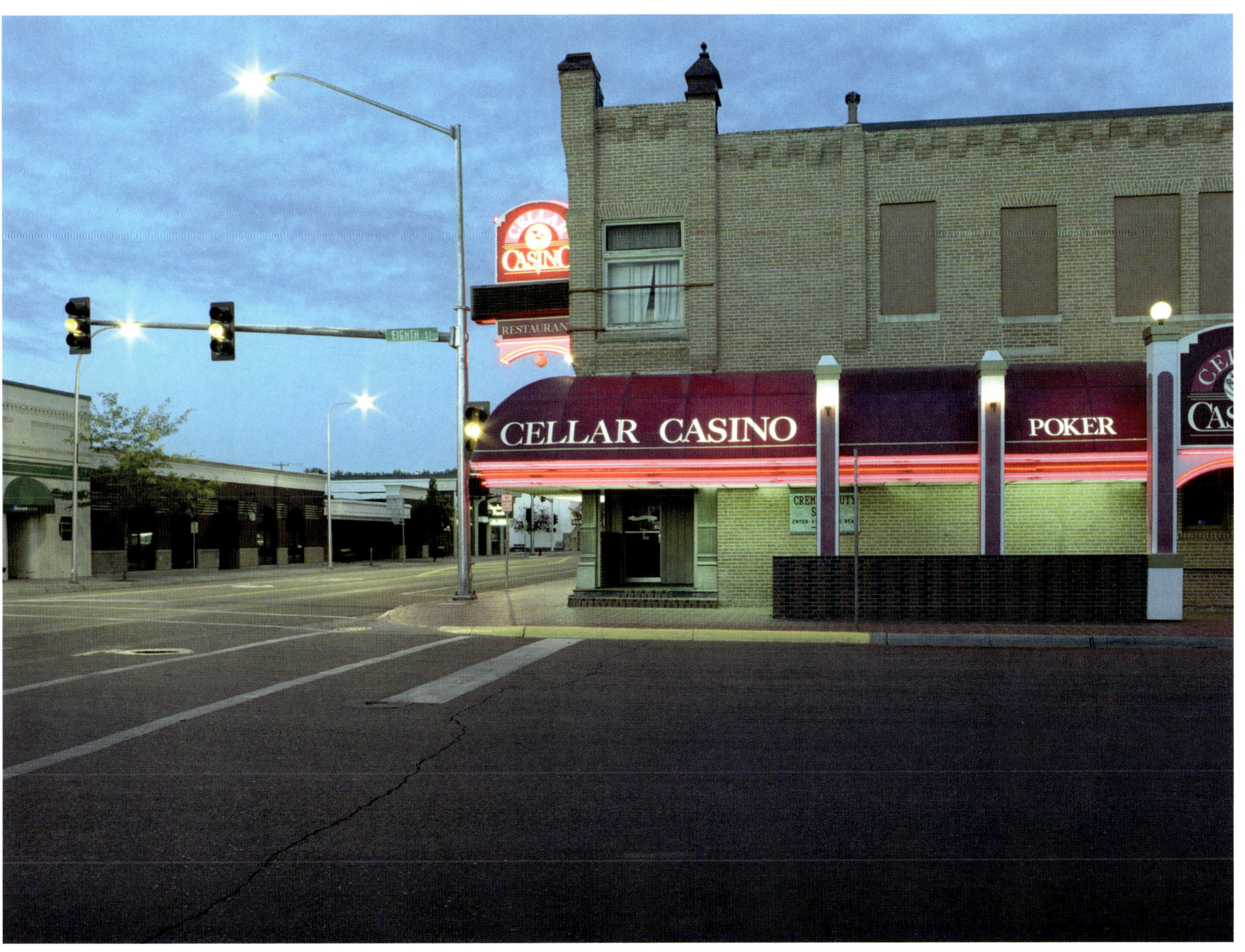

The Cellar Casino, Miles City, Montana, photographed at dawn

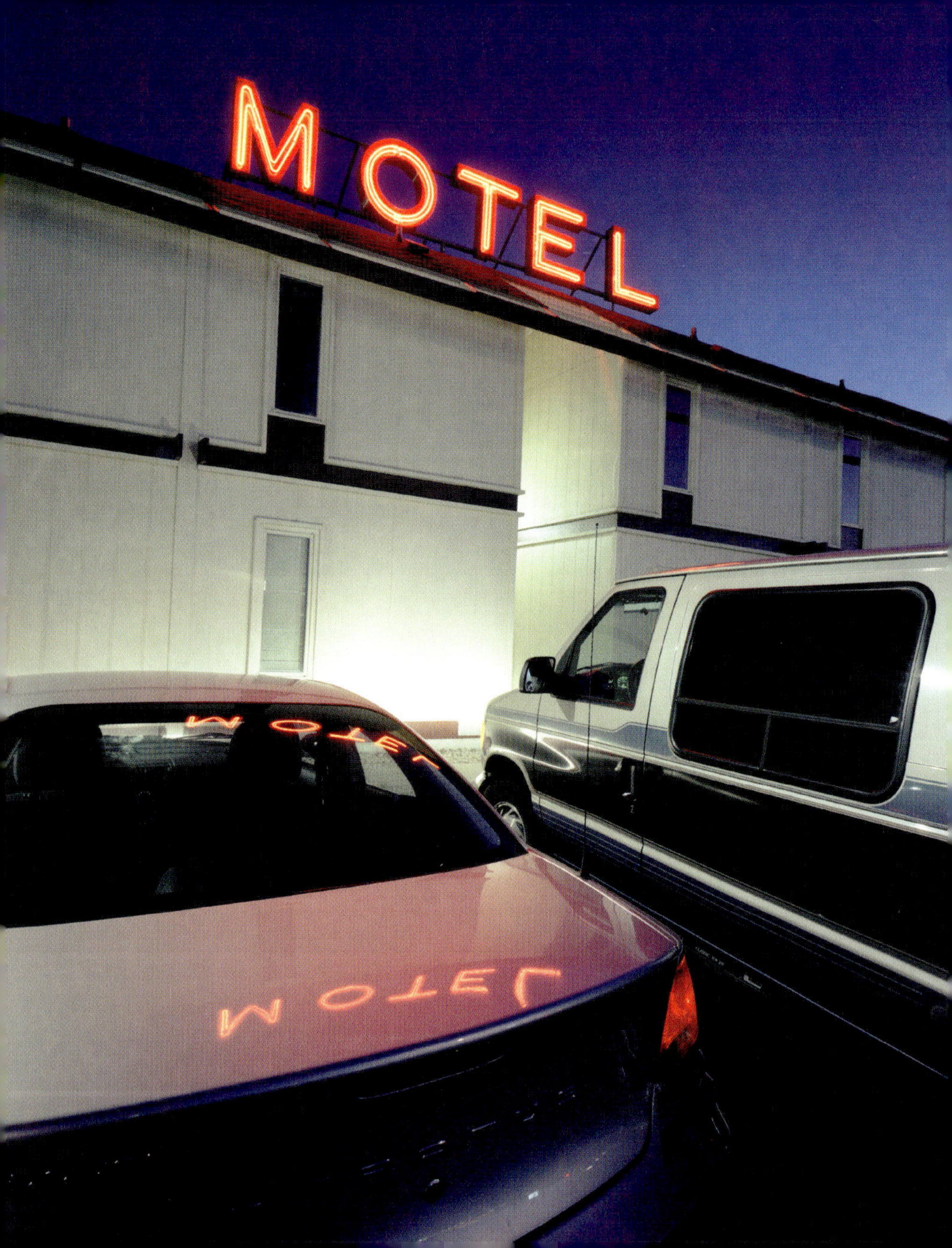
MOTEL

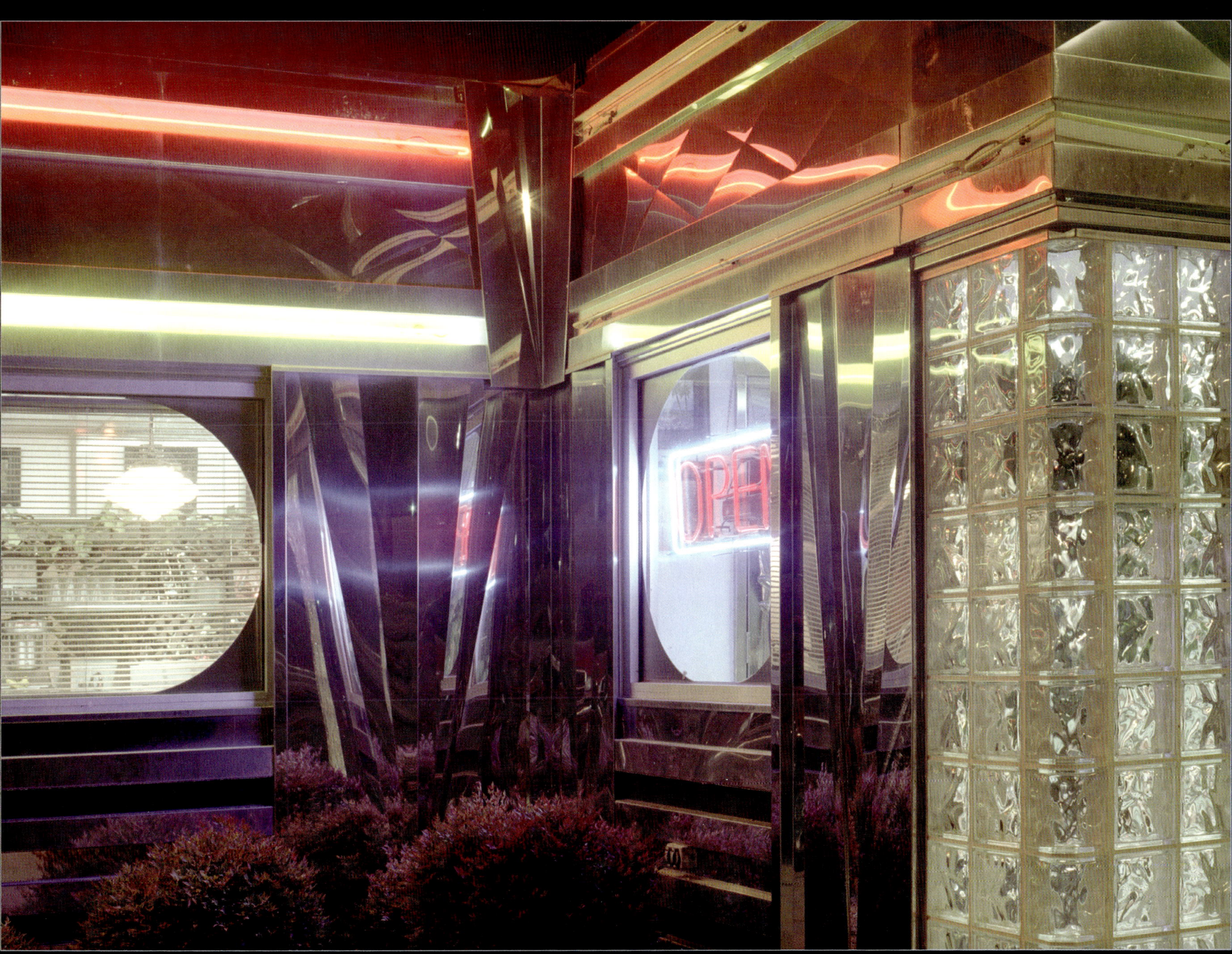
OPEN

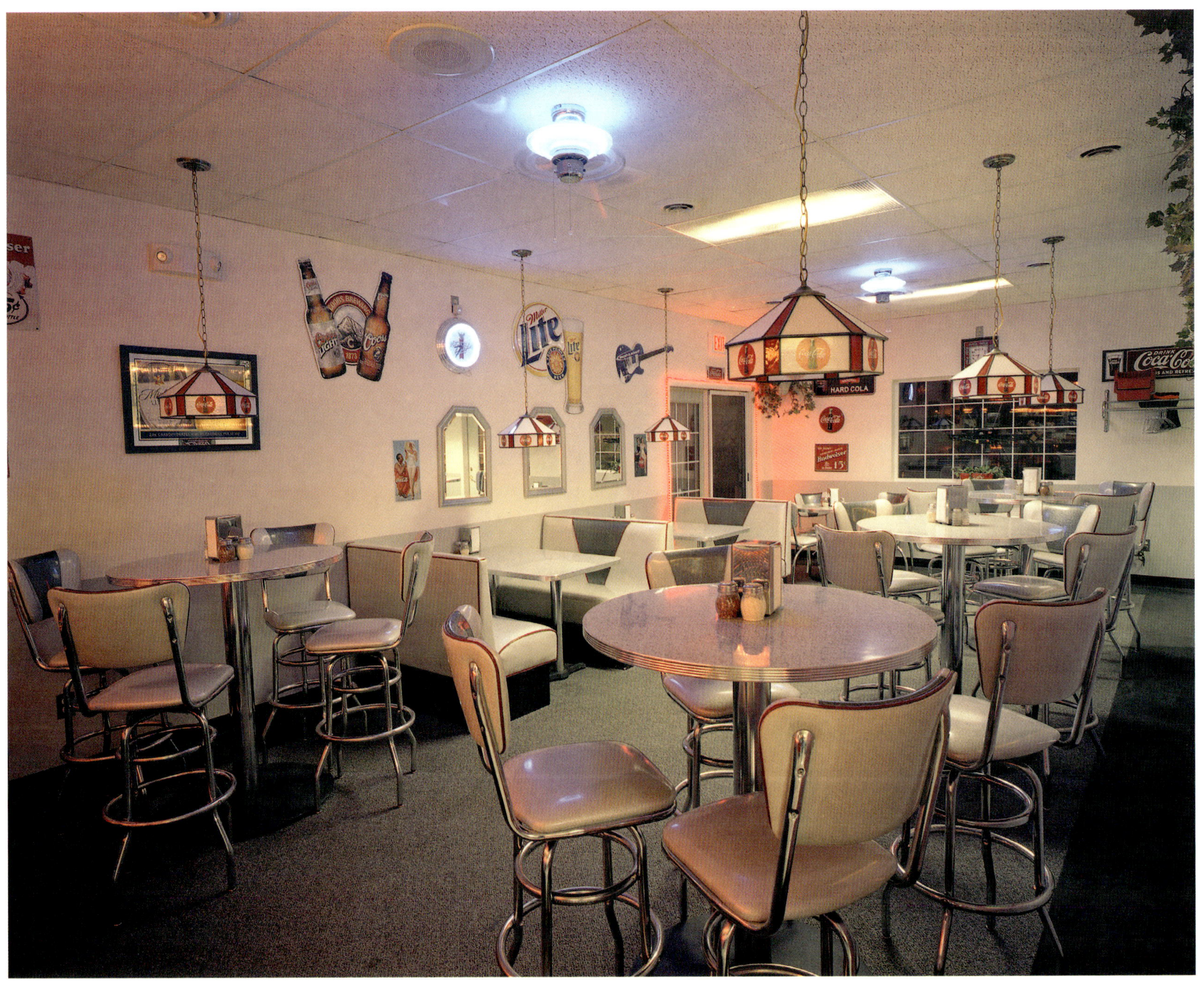

Interior of "Hot Stuff Pizza", Crosby , North Dakota

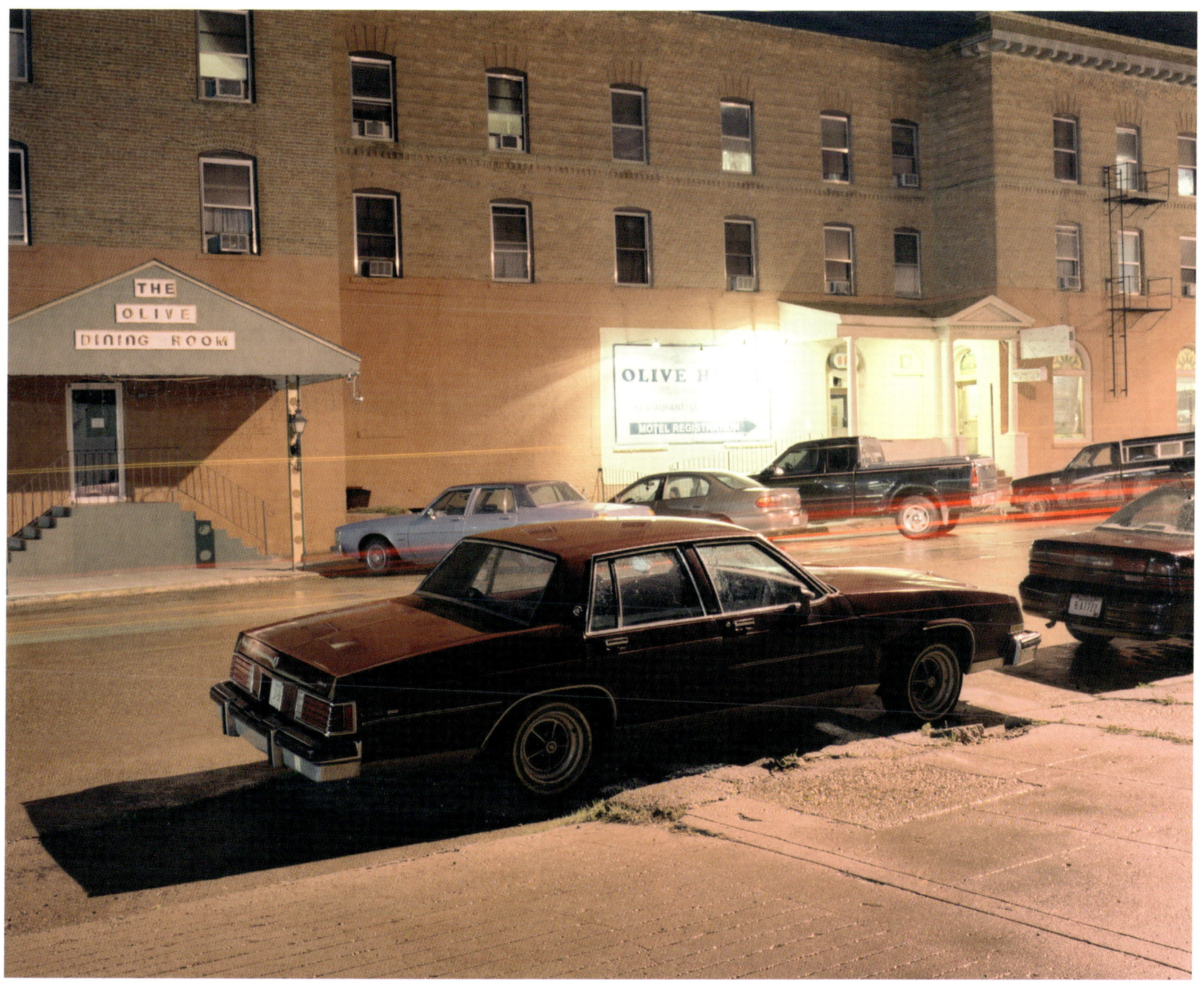

The famous Olive Hotel in Miles City, Montana
Even fictional characters have stayed at this historic location

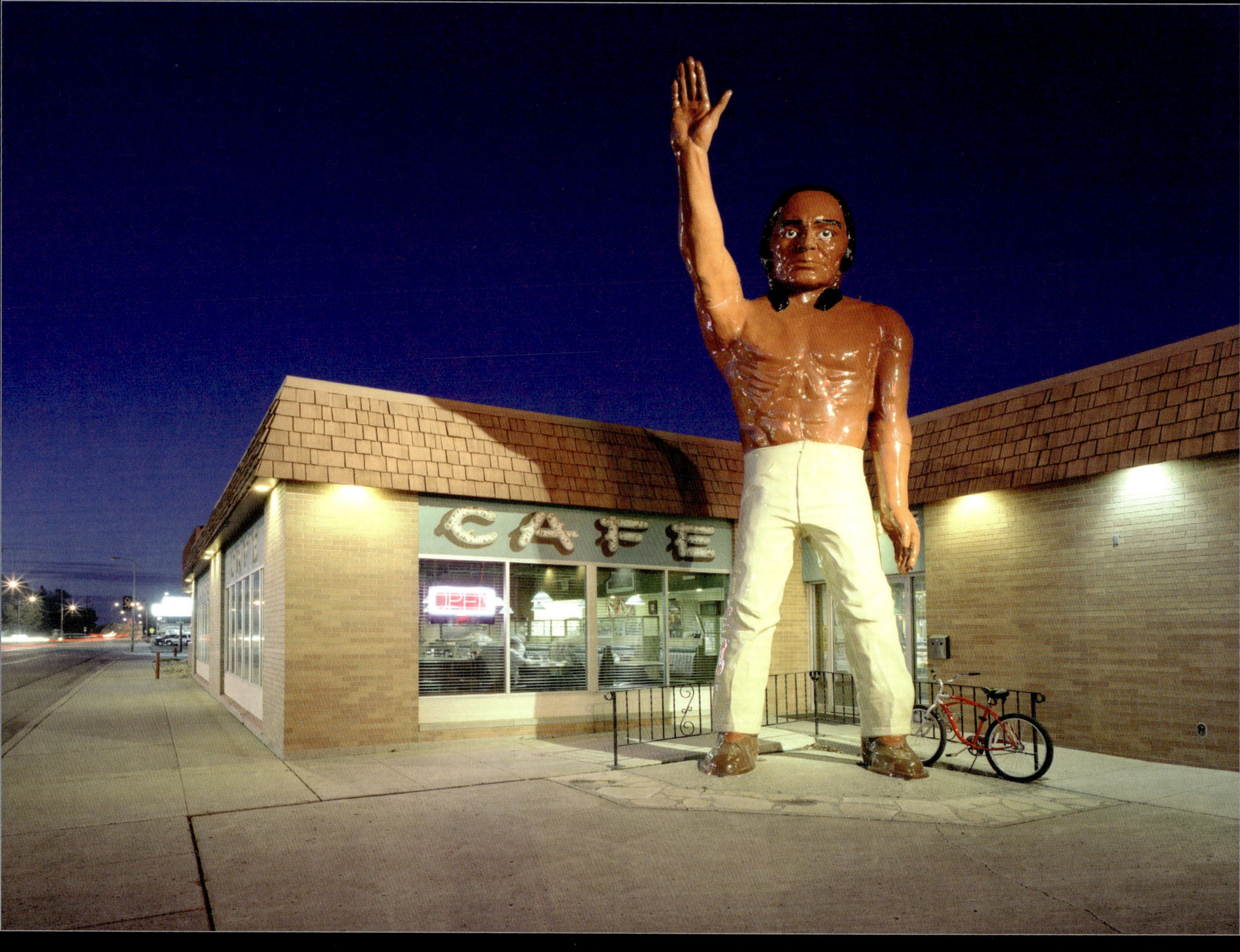

The Chieftain Cafe and Indian Muffler Man, Carrington, North Dakota

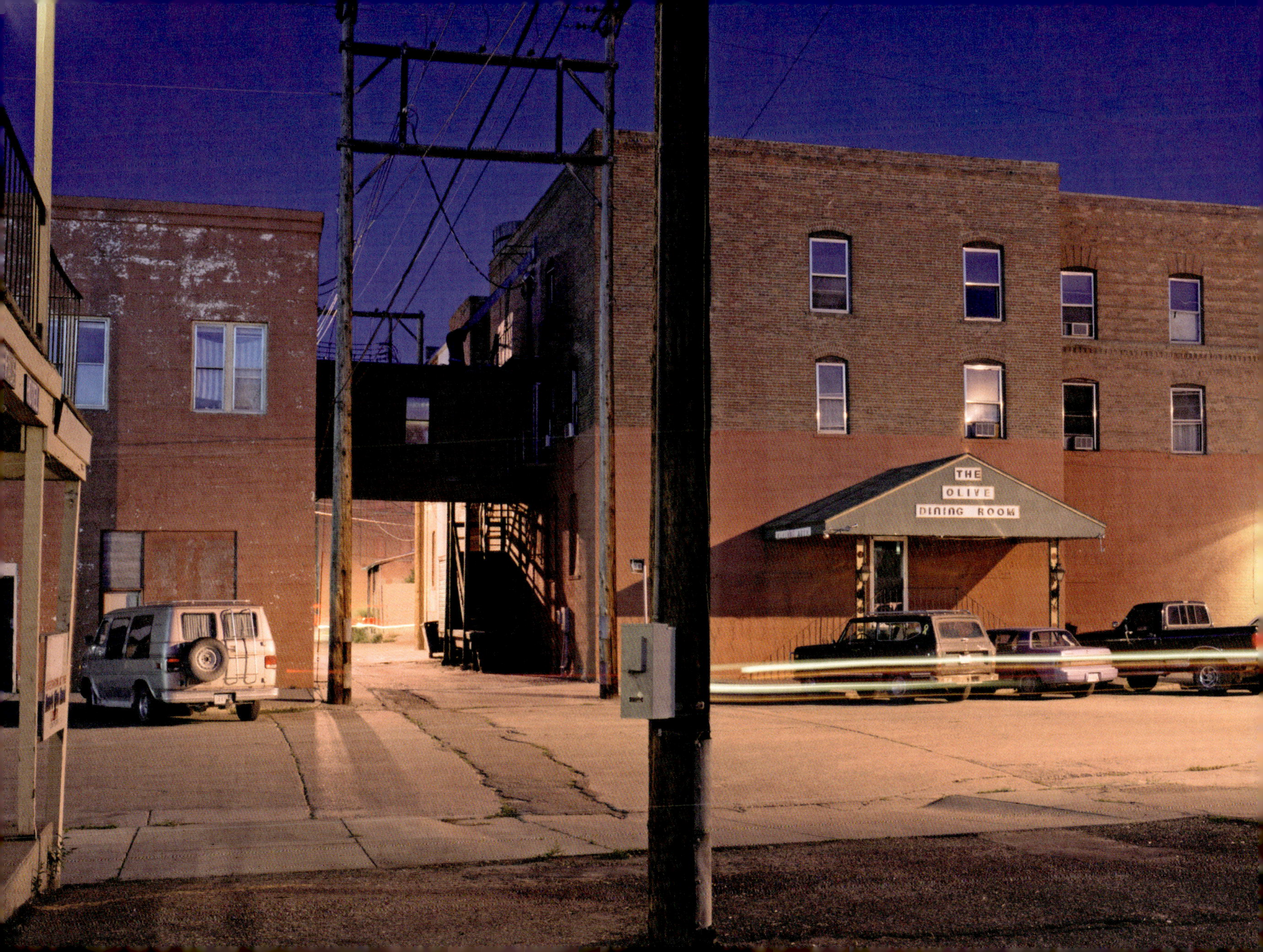

THE
OLIVE
DINING ROOM

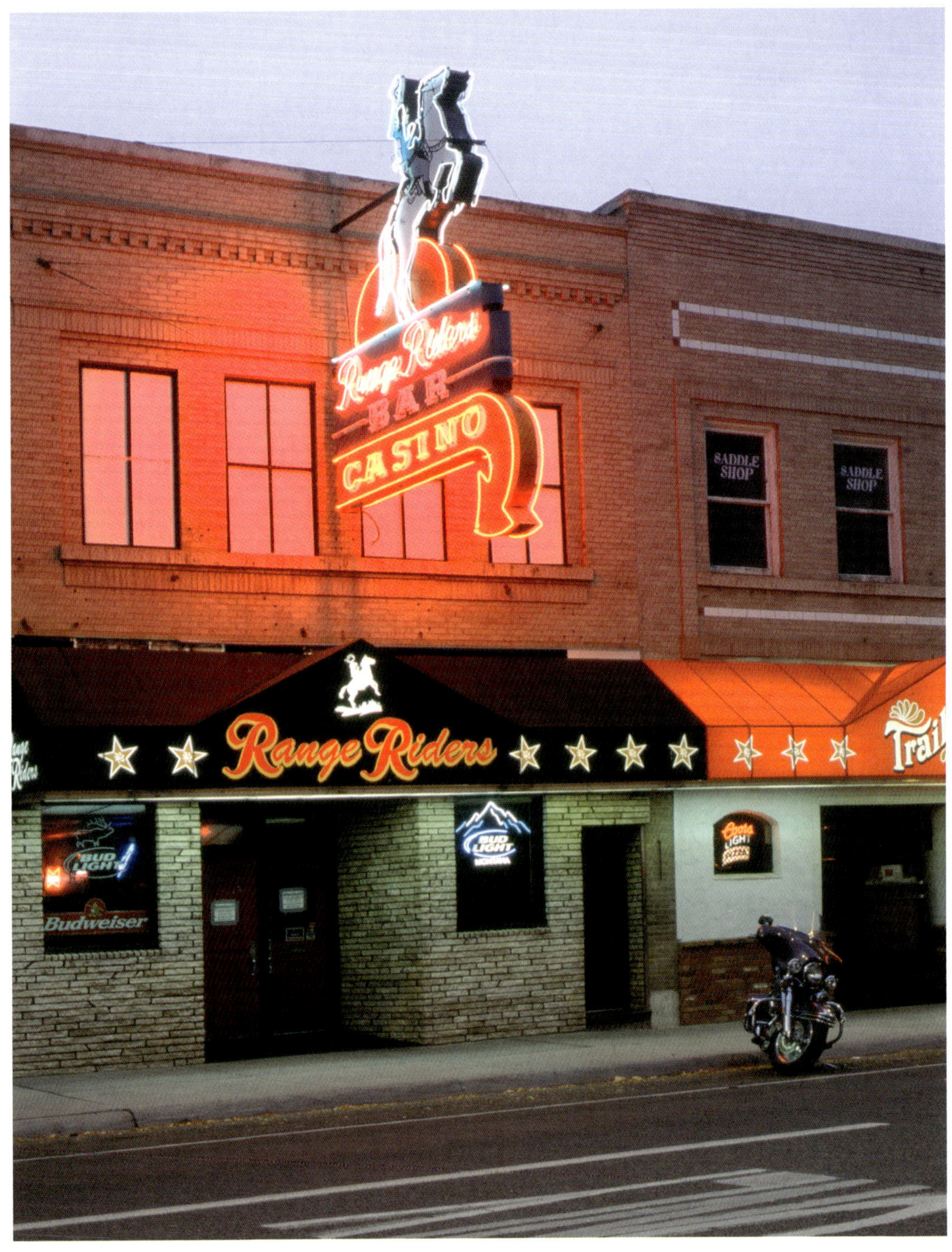

The Range Riders bar, Miles City, Montana

Many of Miles City's atmospheric bars date back to the early 1900s

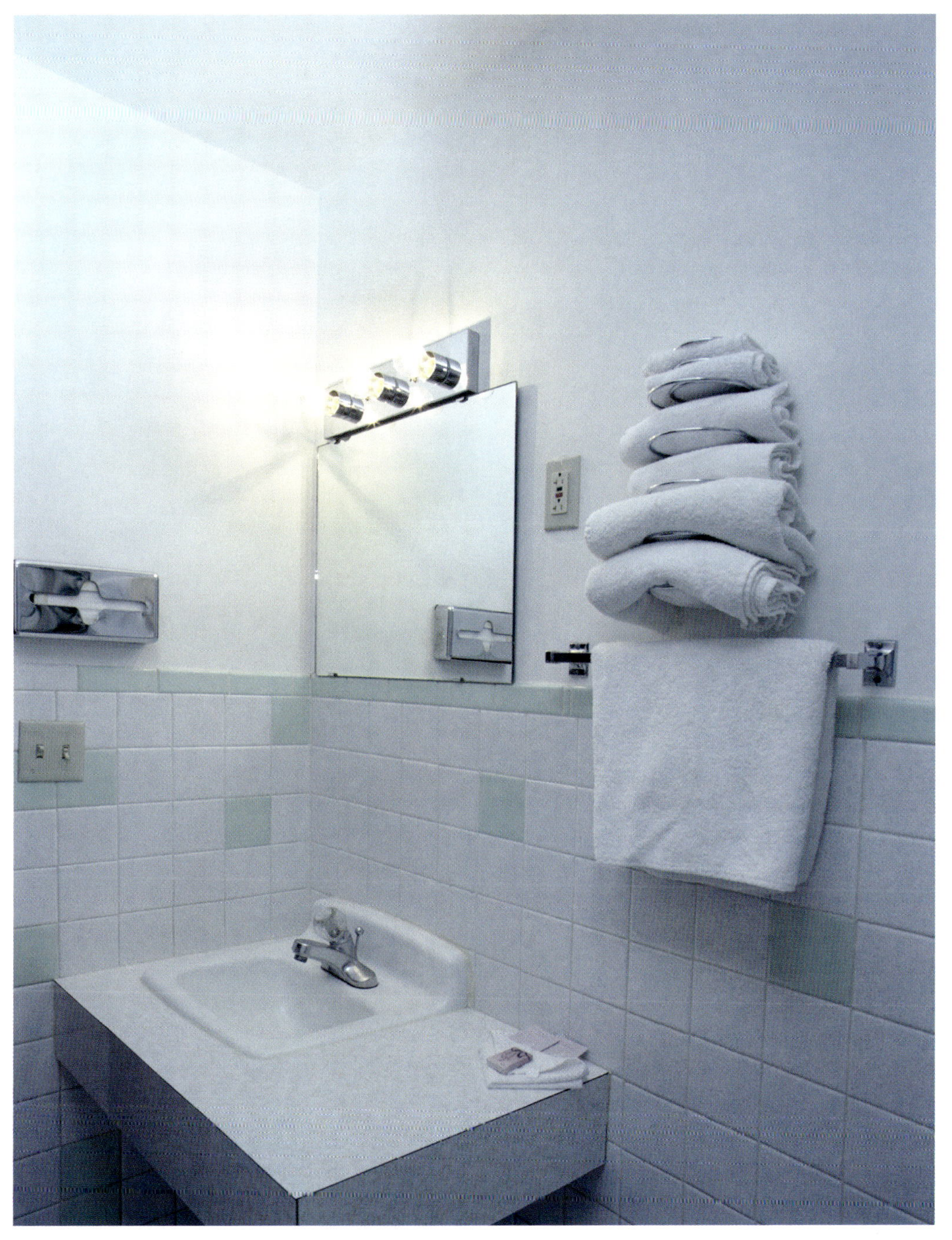

Motel bathroom, Glasgow, Montana

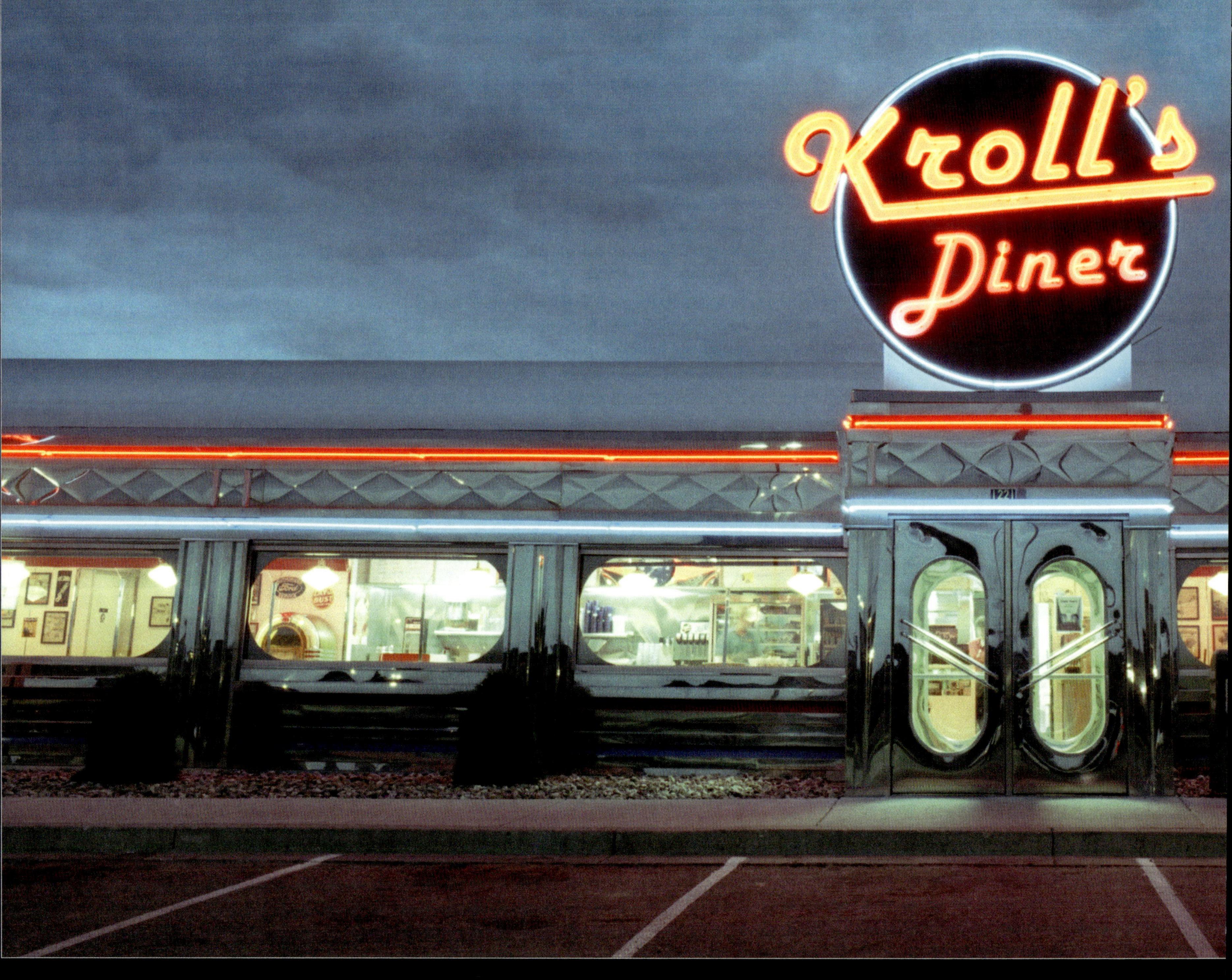

Kroll's
Diner

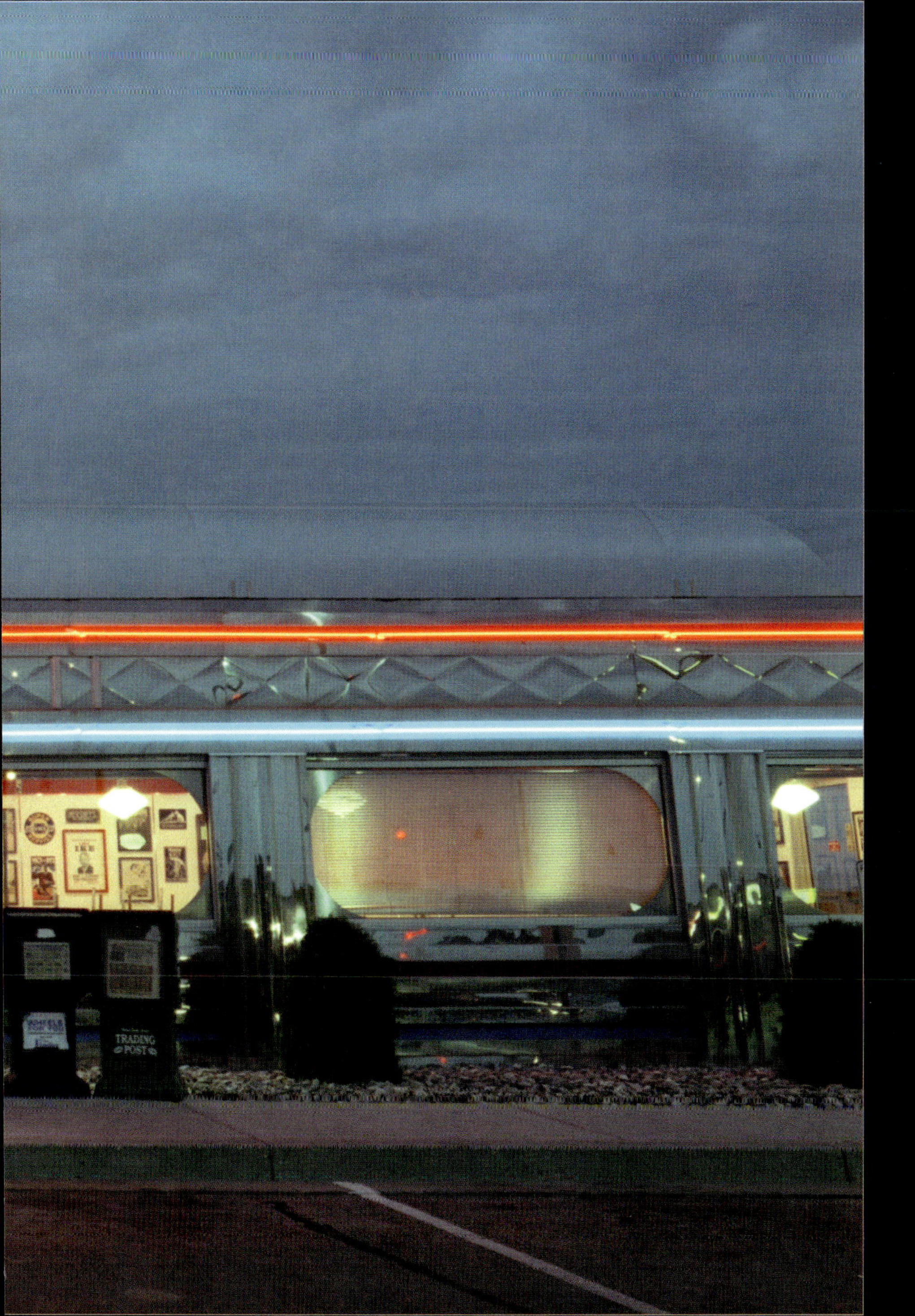

Kroll's Diner, Minot, North Dakota
The interior evokes the atmosphere of the 1950s

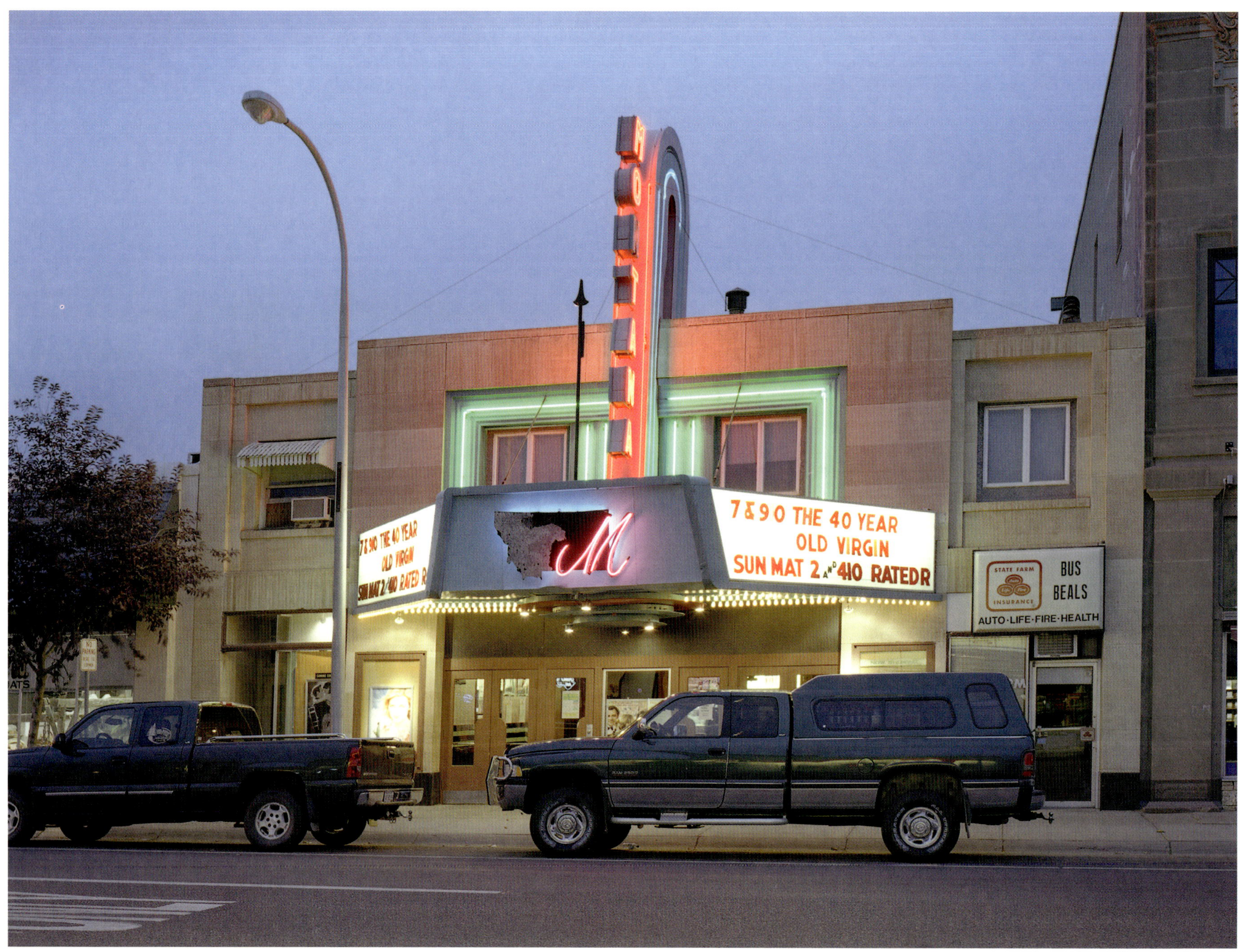

The Montana Theatre, Miles City, Montana
Opened in 1936 and still independently run

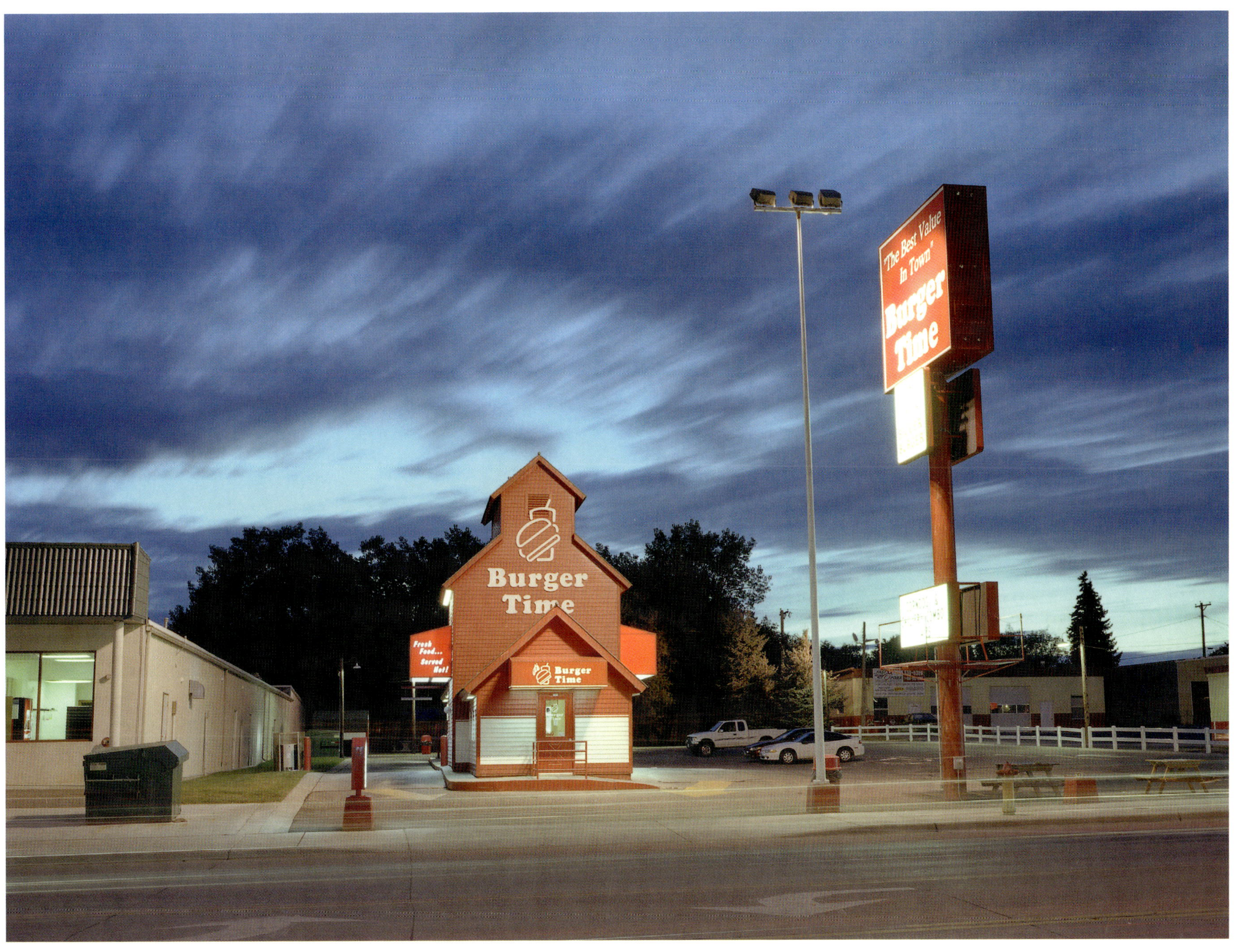

Burger Time, Minot, North Dakota

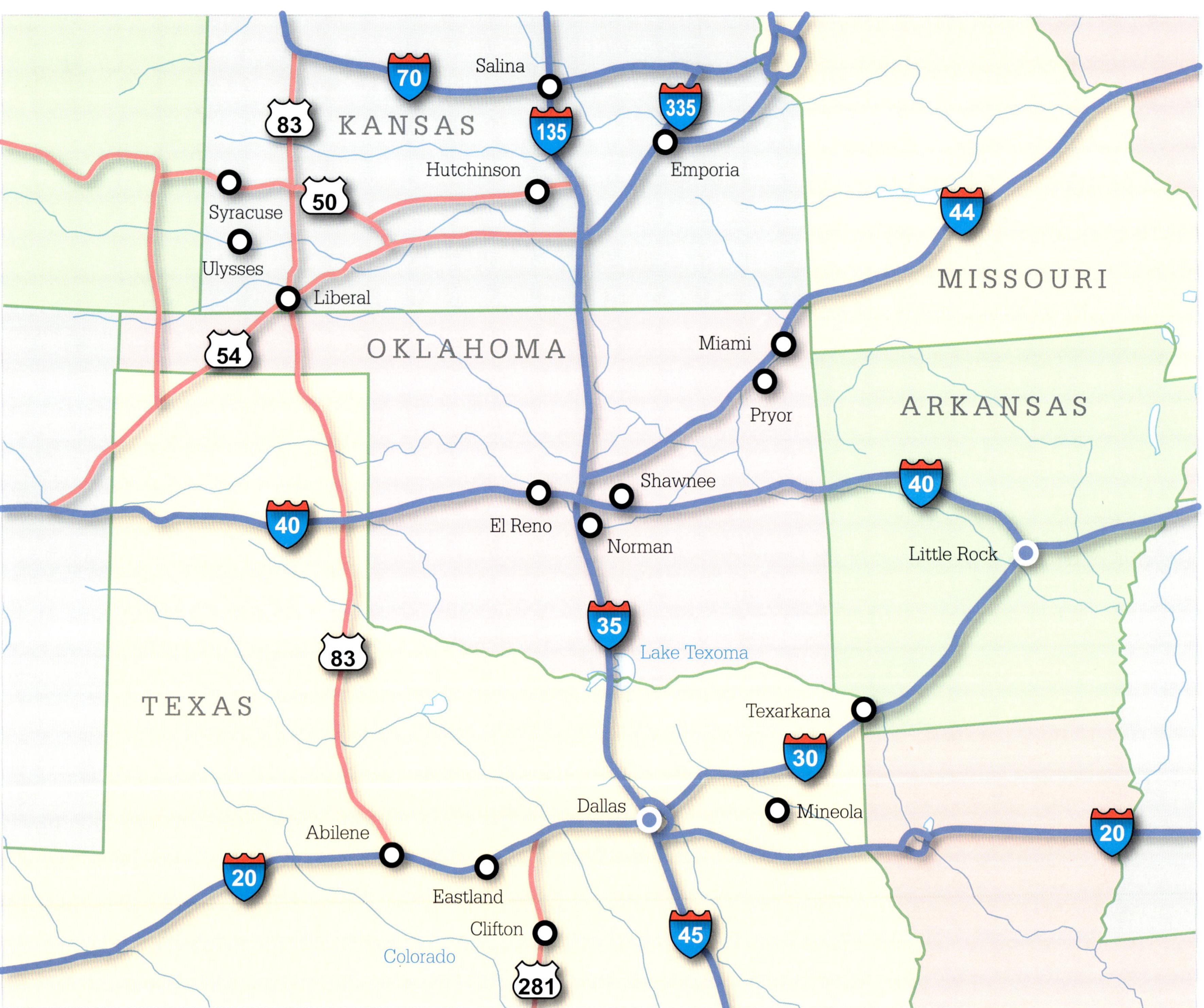
70
Salina
83
135
335
KANSAS
Hutchinson
Emporia
Syracuse
50
44
Ulysses
MISSOURI
Liberal
54
OKLAHOMA
Miami
ARKANSAS
Pryor
Shawnee
40
40
El Reno
Norman
Little Rock
35
Lake Texoma
TEXAS
Texarkana
83
30
Mineola
Dallas
20
Abilene
20
Eastland
Clifton
45
Colorado
281

Follow the Yellow Brick Road

It is sometimes difficult to appreciate just how vast the state of Texas is, stretching as it does from the Gulf of Mexico to the panhandle. Its landscape is often characterized as flat and boring, full of shredded tires, road-kill and hovering vultures, but this stereotype does not do the state justice. Once you are outside the vast metropolis formed by the triangle of Dallas, San Antonio and Houston, you enter a land of contrasts. Expansive ranches where distant cattle dot the horizon do certainly feature, but this is also an area of intensive agriculture, with fields of peaches, pecans, peanuts and cotton. Texas is also a wine-growing area.

Travel to the north or east of Dallas and you can find yourself amid some of the most charming landscape imaginable. Gently undulating, the open plains give way to deciduous woodland. If you're lucky enough to be here in spring you'll encounter a rich variety of flora. Throughout the months of March to May, the fields and verges are carpeted in sweet-william, snapdragon and blue-bonnet. The residences in this particularly charmed corner of Texas are as prosperous as any in the more desirable enclaves of America; even the trailer homes look half-decent when set in such salubrious surroundings! It's no surprise that road-names like Wisteria Highway, Rose Avenue and Azalea Lane abound. This is also an area of countless small lakes, making it a magnet for fishermen and water-sport enthusiasts.

Irrespective of the surrounding landscape, most small Texan towns retain an especial magic. Typically, downtown is constructed around an impressive courthouse square. As you approach, there is little need to look for direction signs, as often the courthouse can be seen from miles away. The center has all the businesses any modestly sized community needs, including a hardware store, a barber shop, a cafe and of course the cinema.

Leaving Texas and entering south central Oklahoma near Ardmore, you find the landscape still reflects features of the southern plains. The area is agreeably arable with attractive fields of peach trees and imposing properties scattered across the land – a far cry from the "Okie" reputation this state still suffers. Images conjured from John Steinbeck's *The Grapes of Wrath* (1939) or Dorothea Lange's portraits of the unthinkable hardships of the 1930s are quickly dispelled. Heading north toward Norman, the route takes you past shimmering lakes, natural spas and hillsides that are dappled with fresh green foliage in springtime and intense autumn colors in the fall. As you continue northeast for a while, you find place-names like Shawnee, Okmulgee and Muskogee that bear witness to Oklahoma's strong Native American heritage, while small towns such as Pryor and Miami manage to maintain vitality in their central business districts through conscious efforts by the citizenry to preserve the local movie theaters.

Over the border in Kansas you come across pleasant communities like Coffeyville, Cherryville and Chanute. The caricatures of this state, complete with folksy gingham pinafores and the like, might be slightly unkind, but many regard Kansas as the true Middle America; if any state were likely to preserve its small towns, surely it would have to be this one. The landscape is undoubtedly verdant and prosperous, although the words "pancake" and "flat" spring readily to mind. The roads are as straight as laser beams, occasionally passing white clapperboard farmhouses reminiscent of the Little House on the Prairie. Kansas is also home to countless Mennonite communities, no doubt attracted by the fertile pastures, civic decency and respect for privacy.

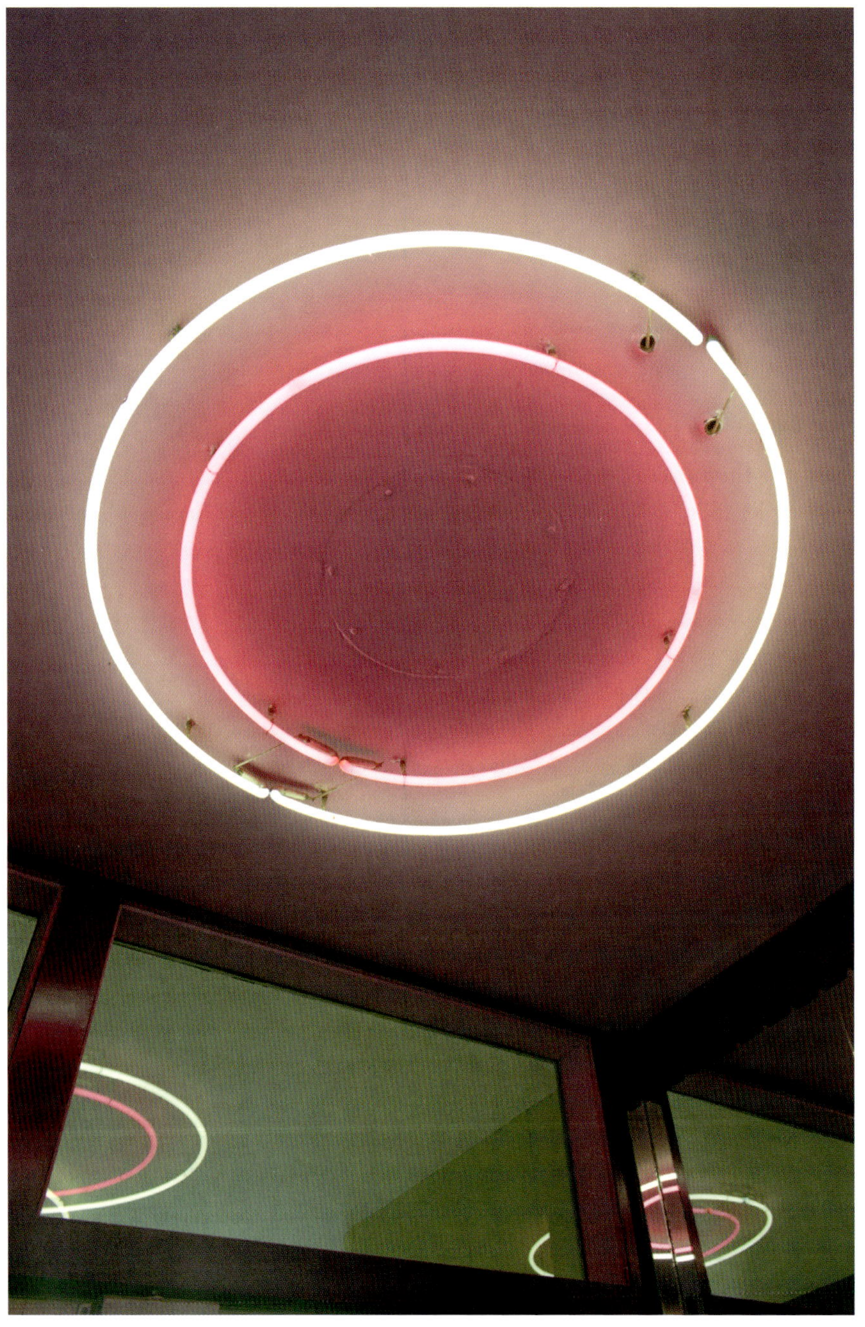

Foyer, "The Movies", Ulysses, Kansas

The towns remain as charming as ever, particularly Iola, with its grand downtown buildings and colorful decorative facades. The town of Liberal, in the southwestern corner of Kansas, offers a strange paradox. Situated on the edge of a large oil and gas field, it certainly has a drab commercial side, but it is nevertheless hard not to notice the local associations (an obsession, one might say) with L. Frank Baum's *The Wonderful Wizard of Oz* (1900). Dorothy's house has been preserved and the Yellow Brick Road identified as if they were historical features rather than figments of an author's imagination. Ozmania is rampant and the annual Oztoberfest attracts thousands, although in truth the link between book and town is at best tenuous.

The Great Plains, stretching well over a thousand miles from southern Texas to the Canadian border in North Dakota and for the most part a couple of hundred miles or more in breadth, are viewed by many Americans as the backbone of the country. In the past, routes forged by settlers, cowboys and outlaws criss-crossed this vast area; later came motorists keen only to reach some far-distant destination as quickly as possible. Today, though, more and more people, seeking some respite from the pressures of urban living, are beginning to appreciate the qualities inherent in these rural heartlands. The particularly enchanting area between northern Kansas and central Texas, rich in small towns and tranquil landscapes, eloquently summarizes the values and culture which many Americans are striving to retain.

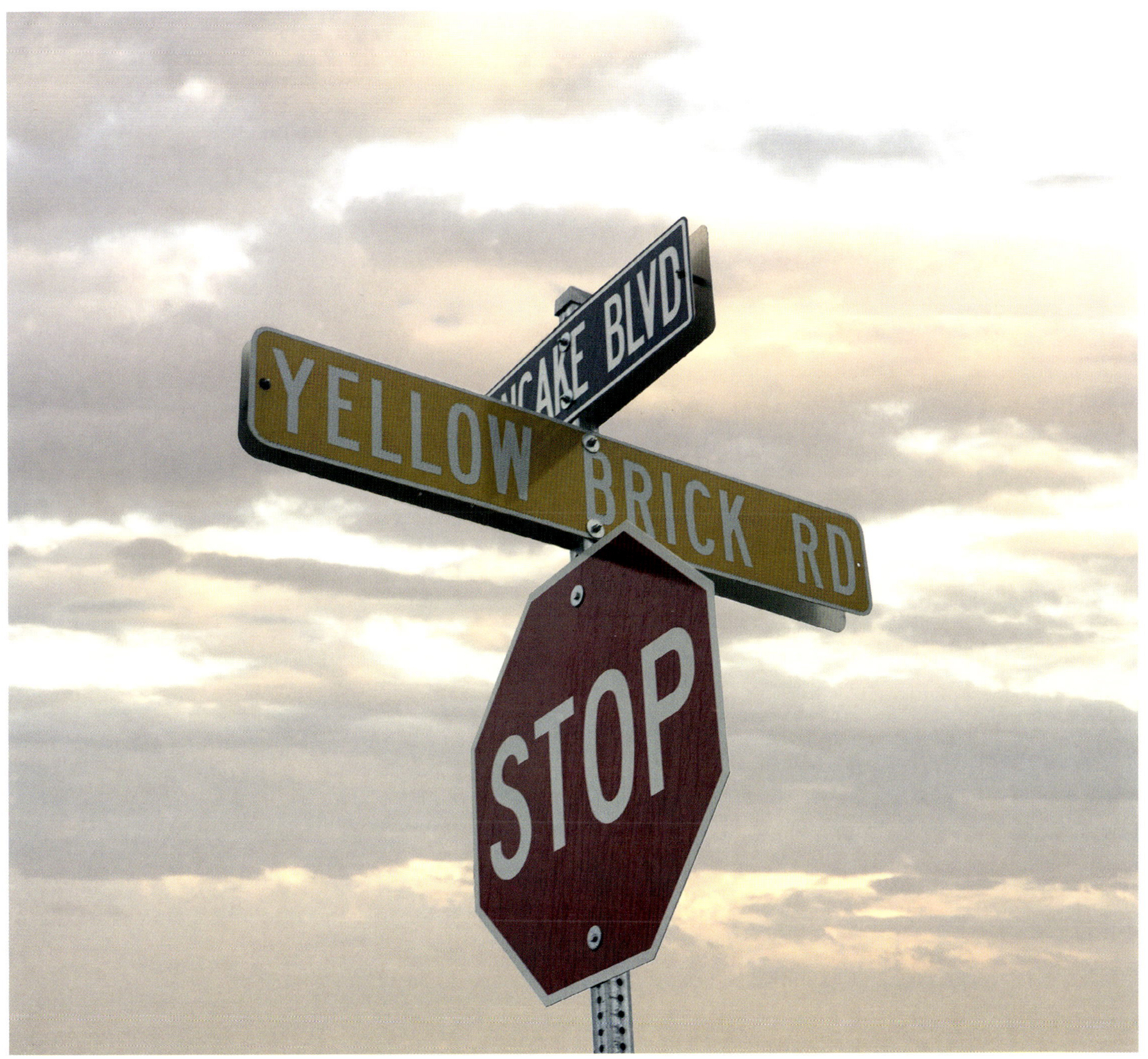

The Yellow Brick Road, Liberal, Kansas

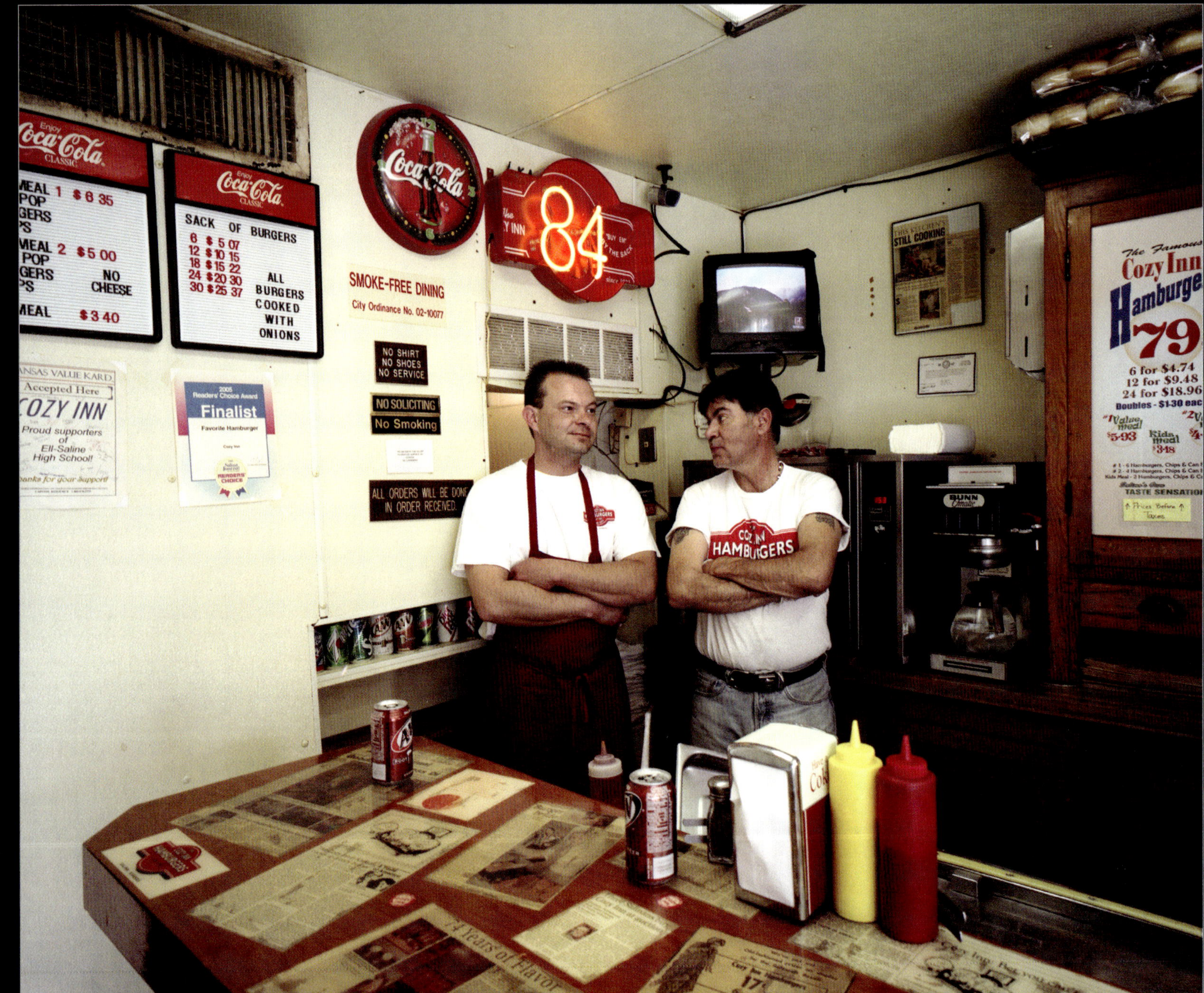
Enjoy
Coca-Cola
CLASSIC
Enjoy
Coca-Cola
CLASSIC
SACK OF BURGERS
6 $ 5 07
12 $10 15
18 $15 22
24 $20 30
30 $25 37
ALL
BURGERS
COOKED
WITH
ONIONS
MEAL 1 $ 6 35
POP
GERS
S
MEAL 2 $5 00
POP
GERS   NO
S   CHEESE
MEAL $3 40
Coca-Cola
84
SMOKE-FREE DINING
City Ordinance No. 02-10077
NO SHIRT
NO SHOES
NO SERVICE
NO SOLICITING
No Smoking
ALL ORDERS WILL BE DONE
IN ORDER RECEIVED.
ANSAS VALUE KARD
Accepted Here
COZY INN
Proud supporters
of
Ell-Saline
High School!
anks for your support
2005
Readers' Choice Award
Finalist
Favorite Hamburger
Cozy Inn
THIS KITCHEN
STILL COOKING
The Famous
Cozy Inn
Hamburger
79
6 for $4.74
12 for $9.48
24 for $18.96
Doubles - $1.30 eac
Value
meal
$5.93
Kids
meal
$3.48
TASTE SENSATION
Prices Before
Taxes.
COZ IN
HAMBURGERS
BUNN

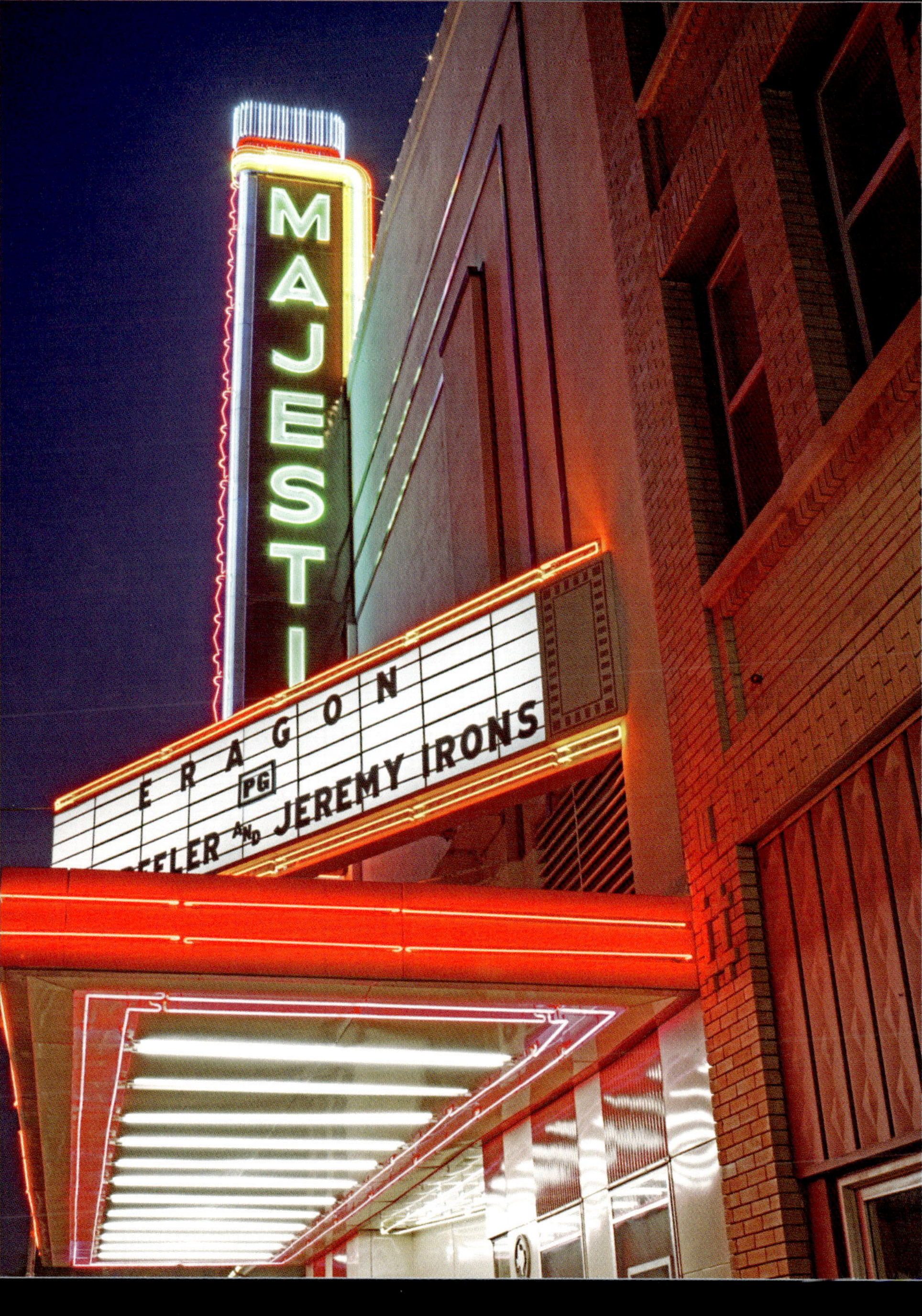

The Majestic Theatre, Eastland, Texas

Built in 1920 and remodeled in 1947 in the Art Moderne style

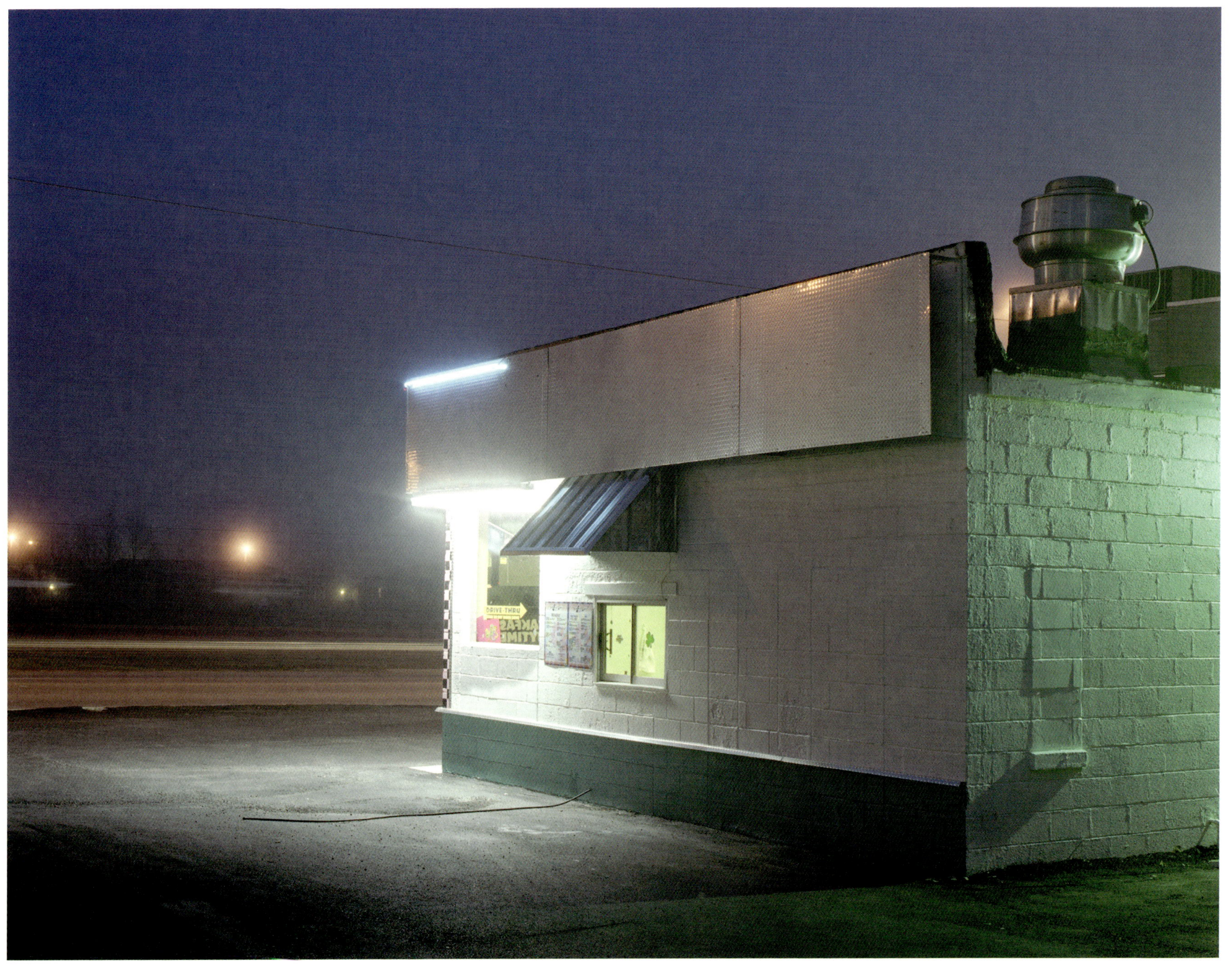

Burger kiosk, Shawnee, Oklahoma

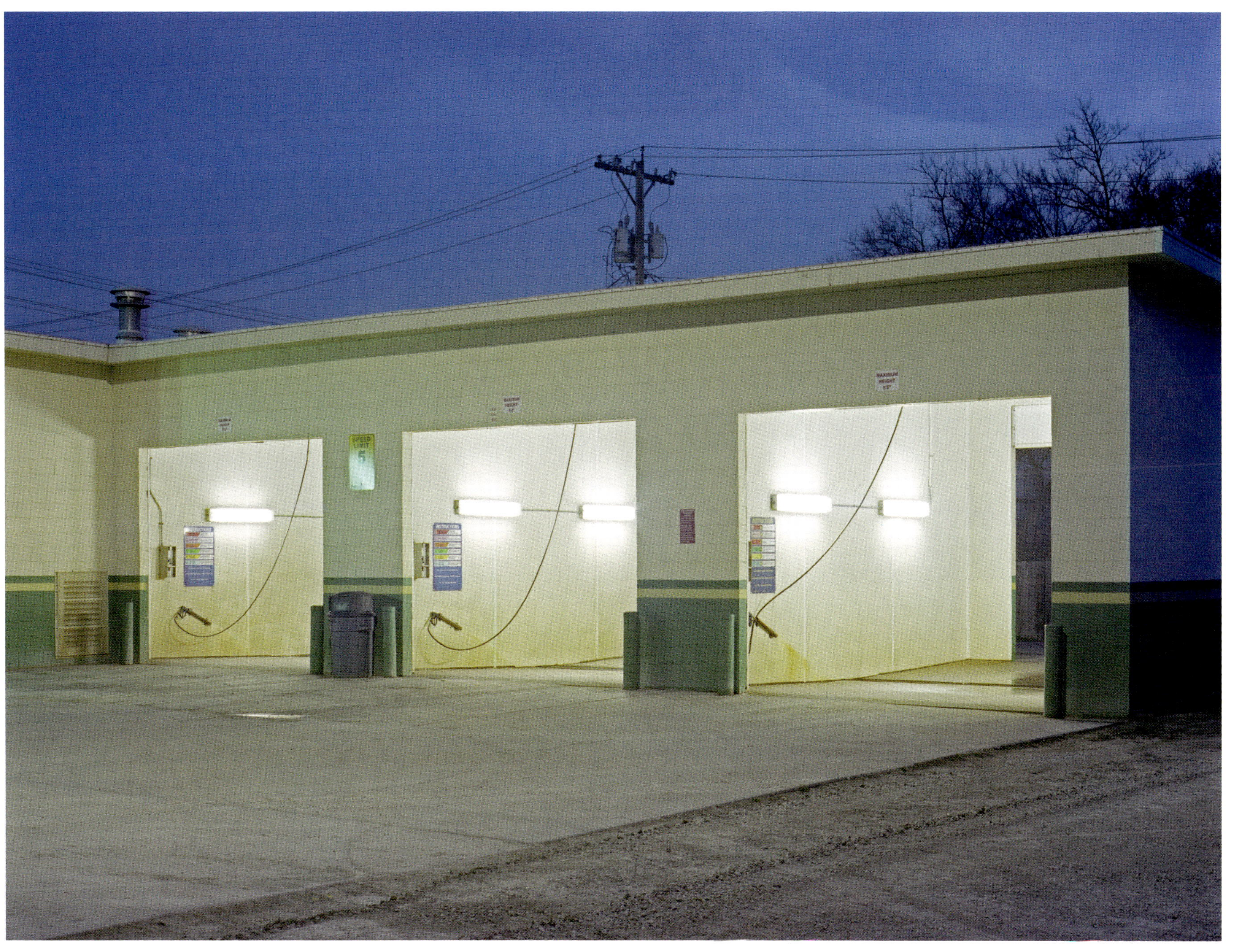

Car wash, Salina, Kansas

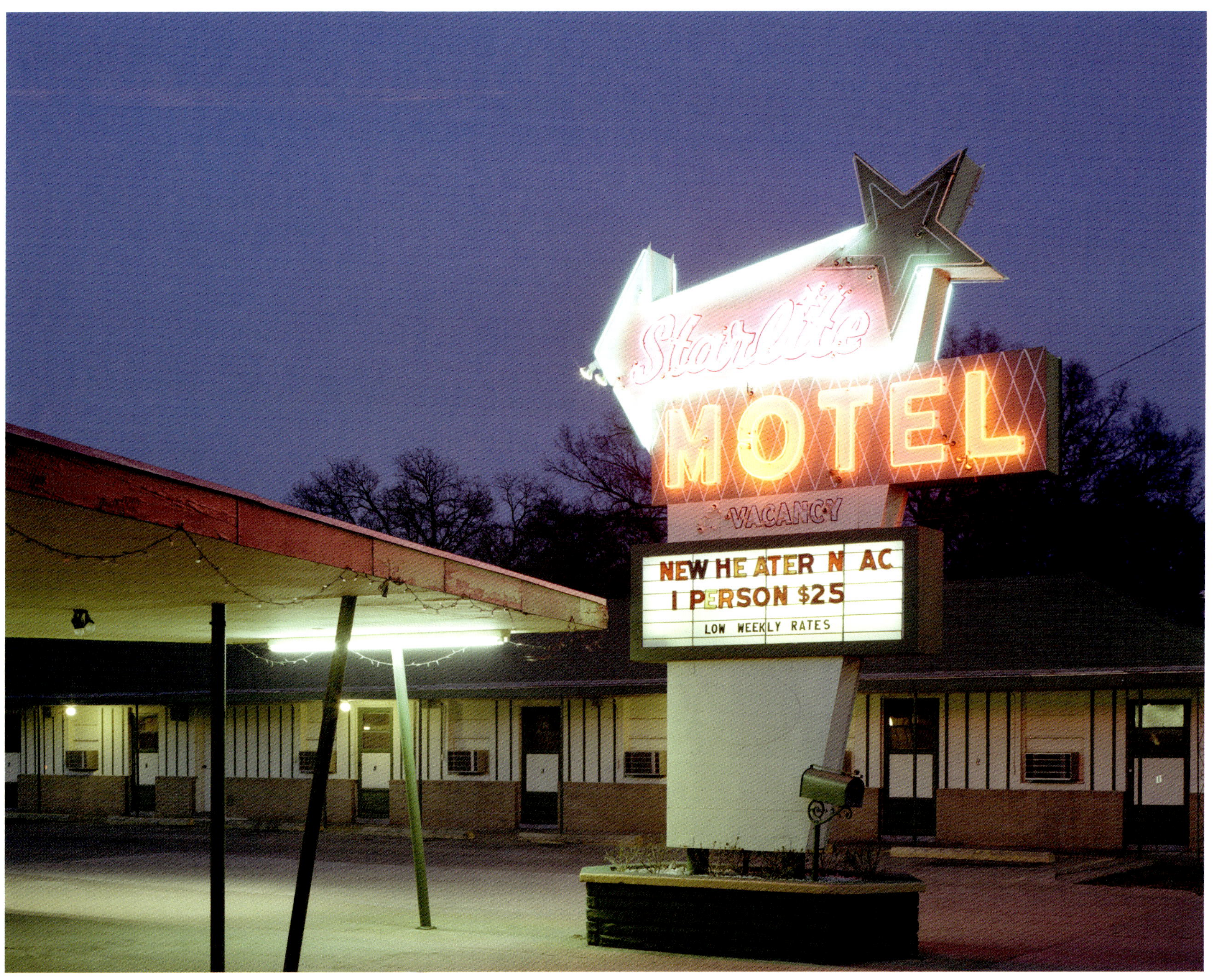

Starlite Motel, Salina, Kansas

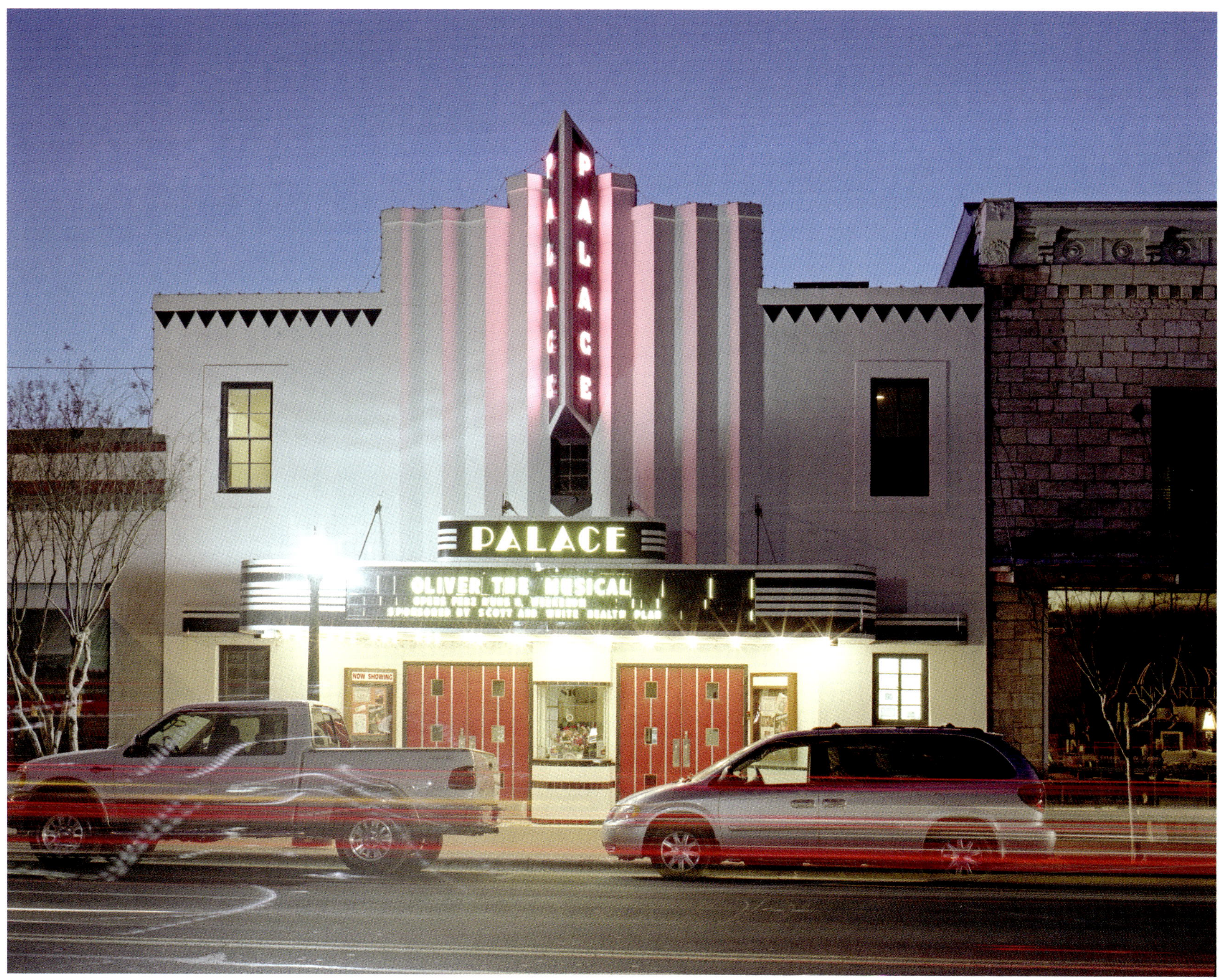

The Palace Theatre, Georgetown, Texas
Originally built in the silent movie era of the 1920s, it was renovated in 1938 as an Art Deco theatre

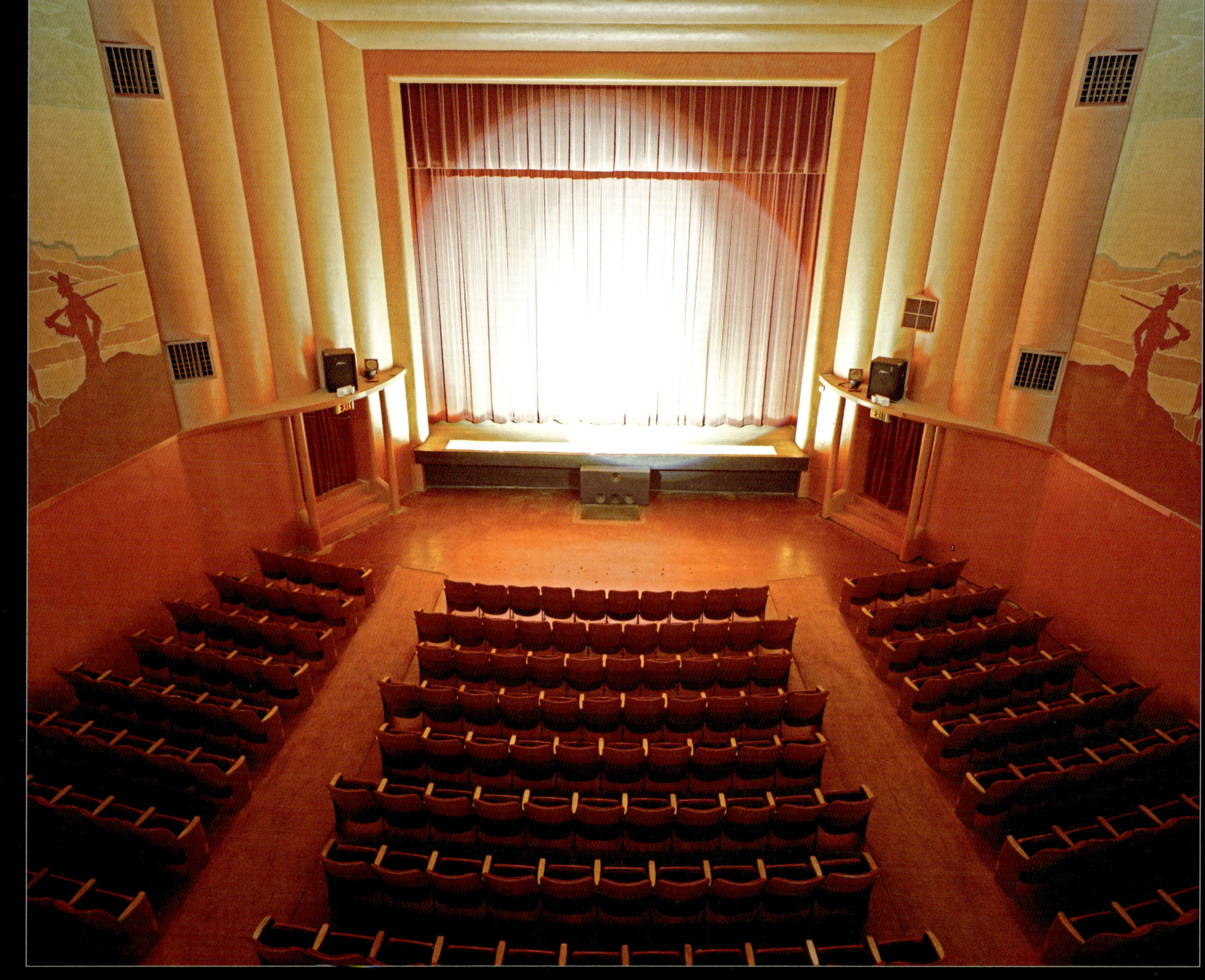

Auditorium, Majestic Theatre, Eastland, Texas

The interiors of such theatres retain a sense of elegance from a bygone era. Gene Autry and his horse appeared on stage here in the 1930s!

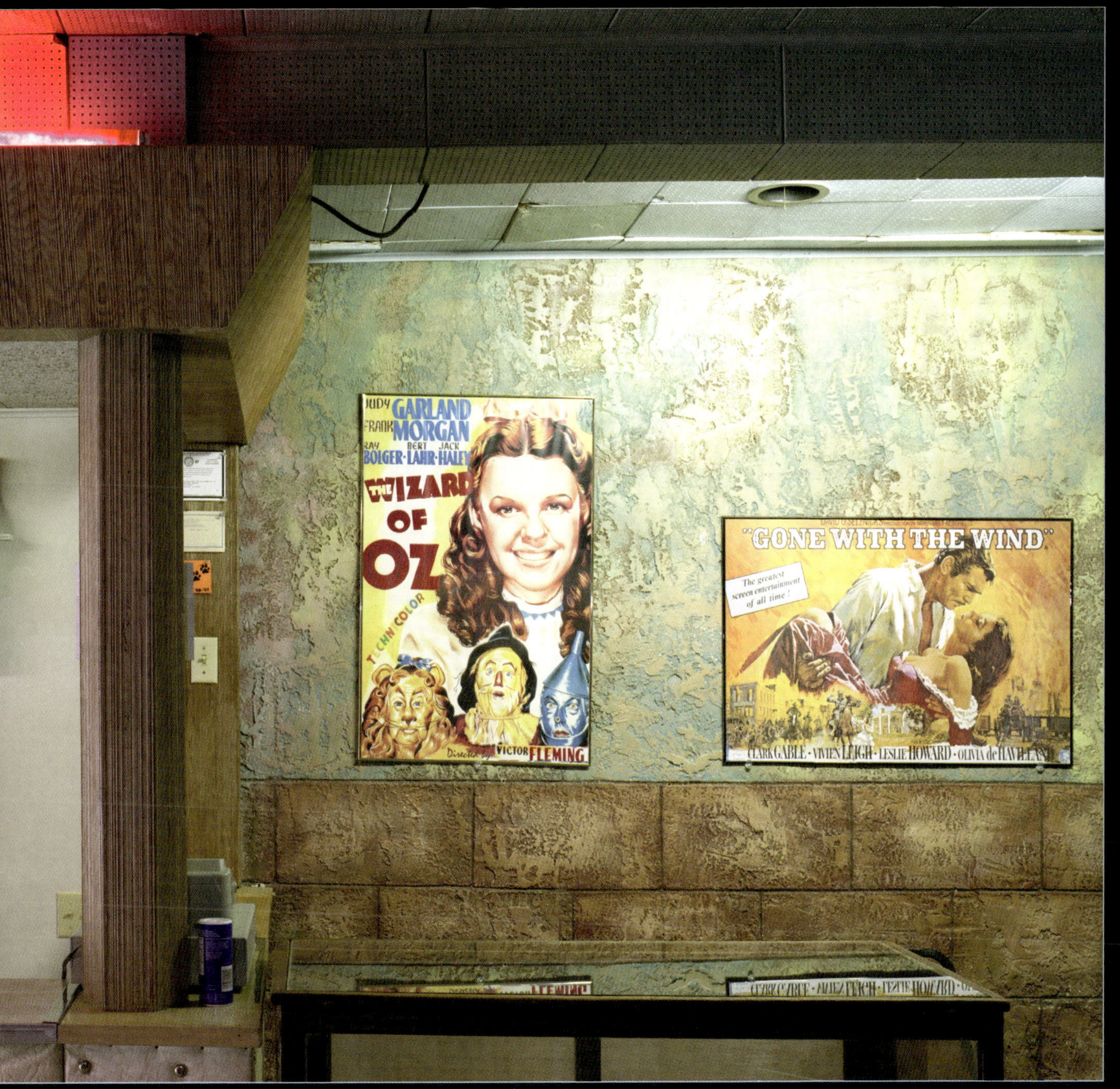

The Wizard of Oz, foyer, Northrup Theatre, Syracuse, Kansas

This Syracuse landmark was added to the National Register of Historic Places in 2005

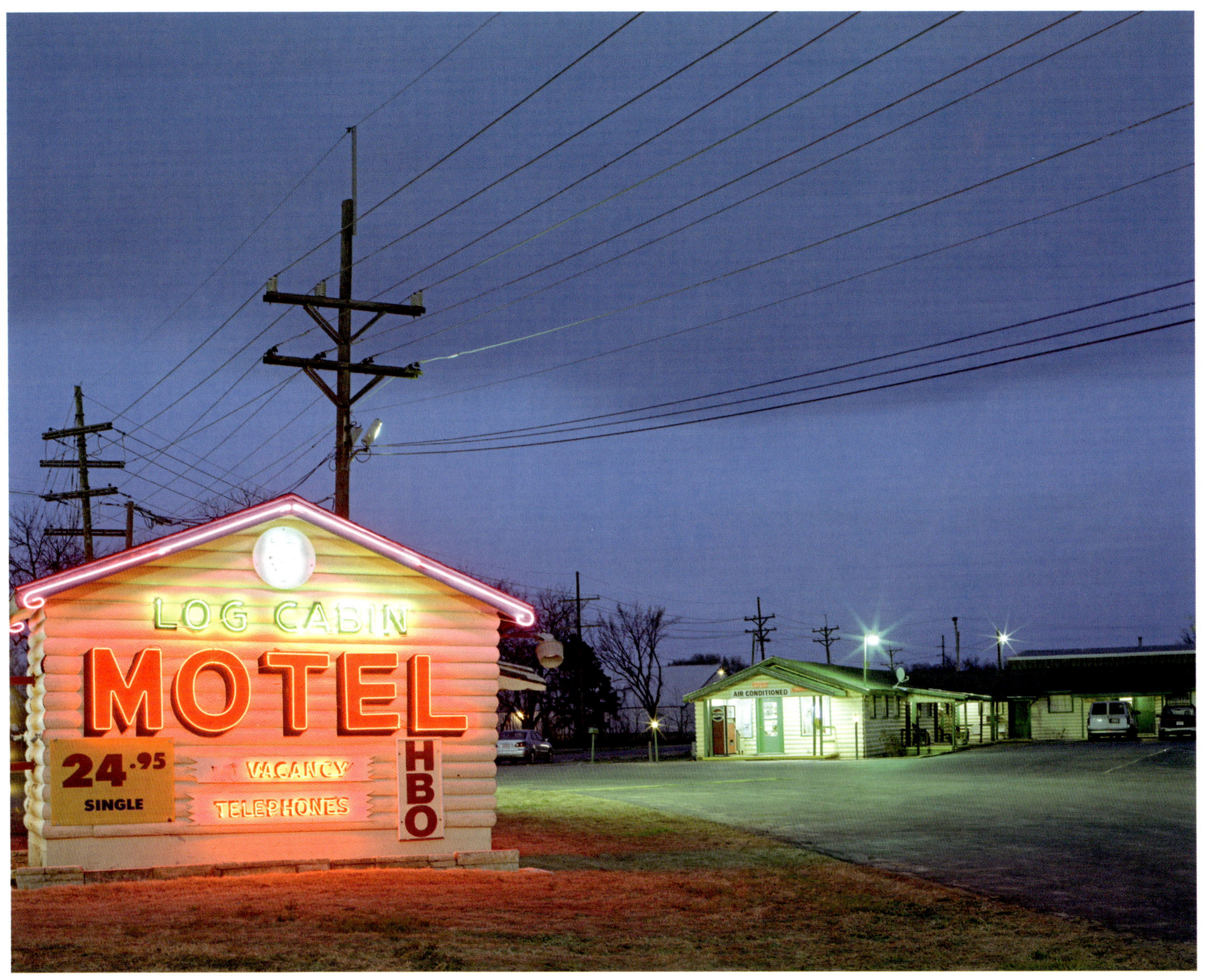

Log Cabin Motel, Salina, Kansas

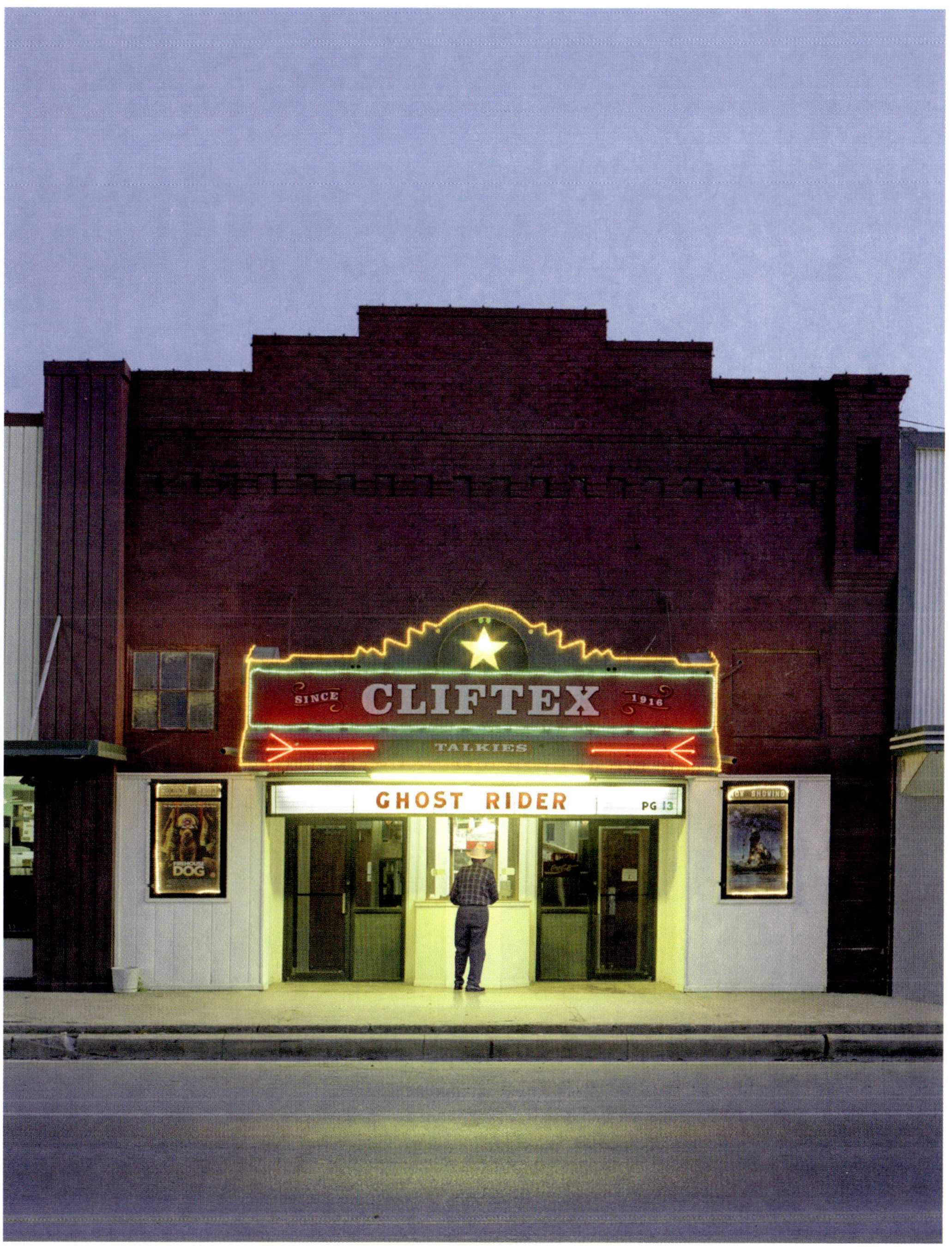

The Cliftex, Clifton, Texas
Opened in 1916, the Cliftex is probably the oldest continually running movie theatre in Texas

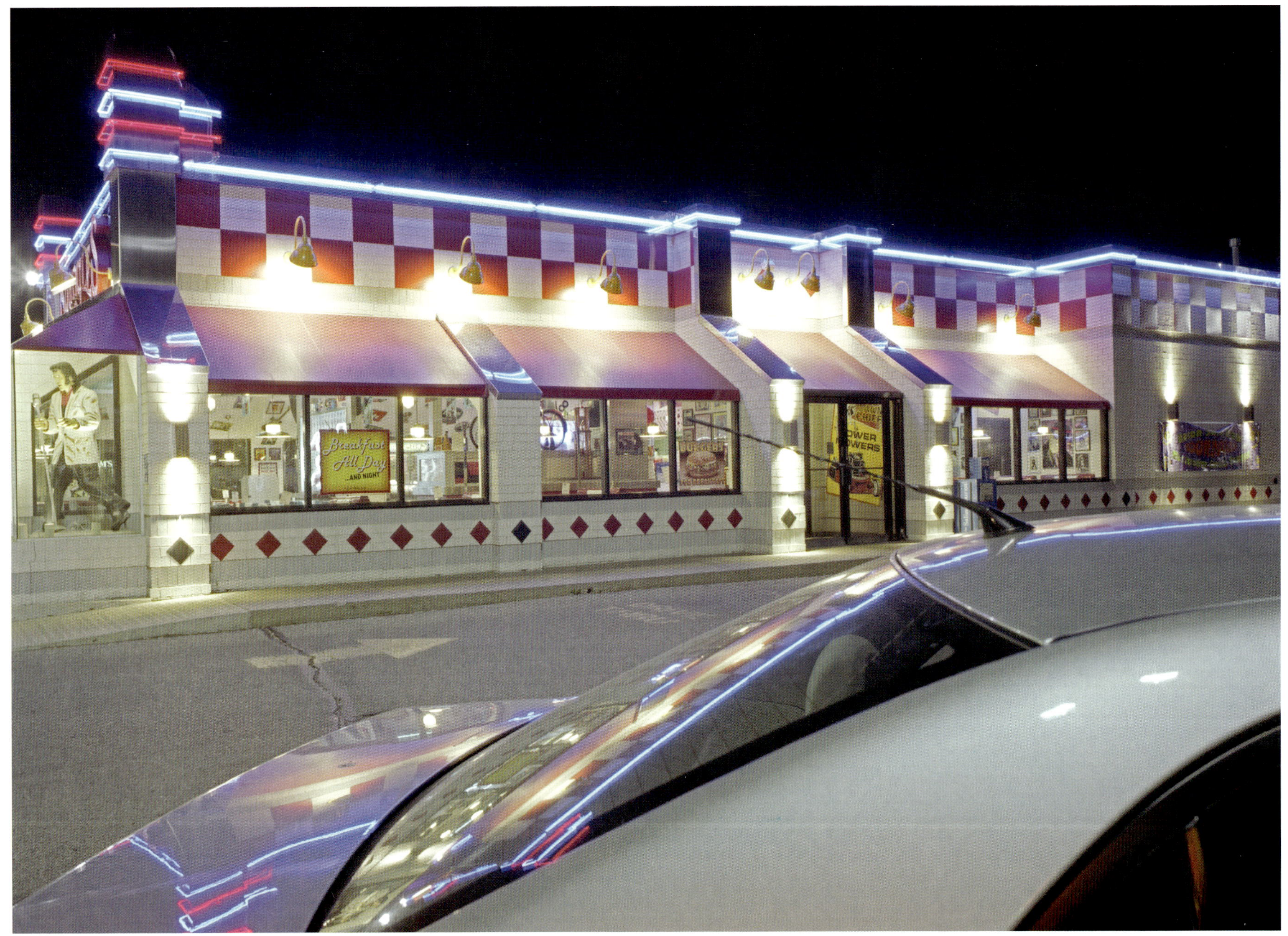

Diner, Salina, Kansas

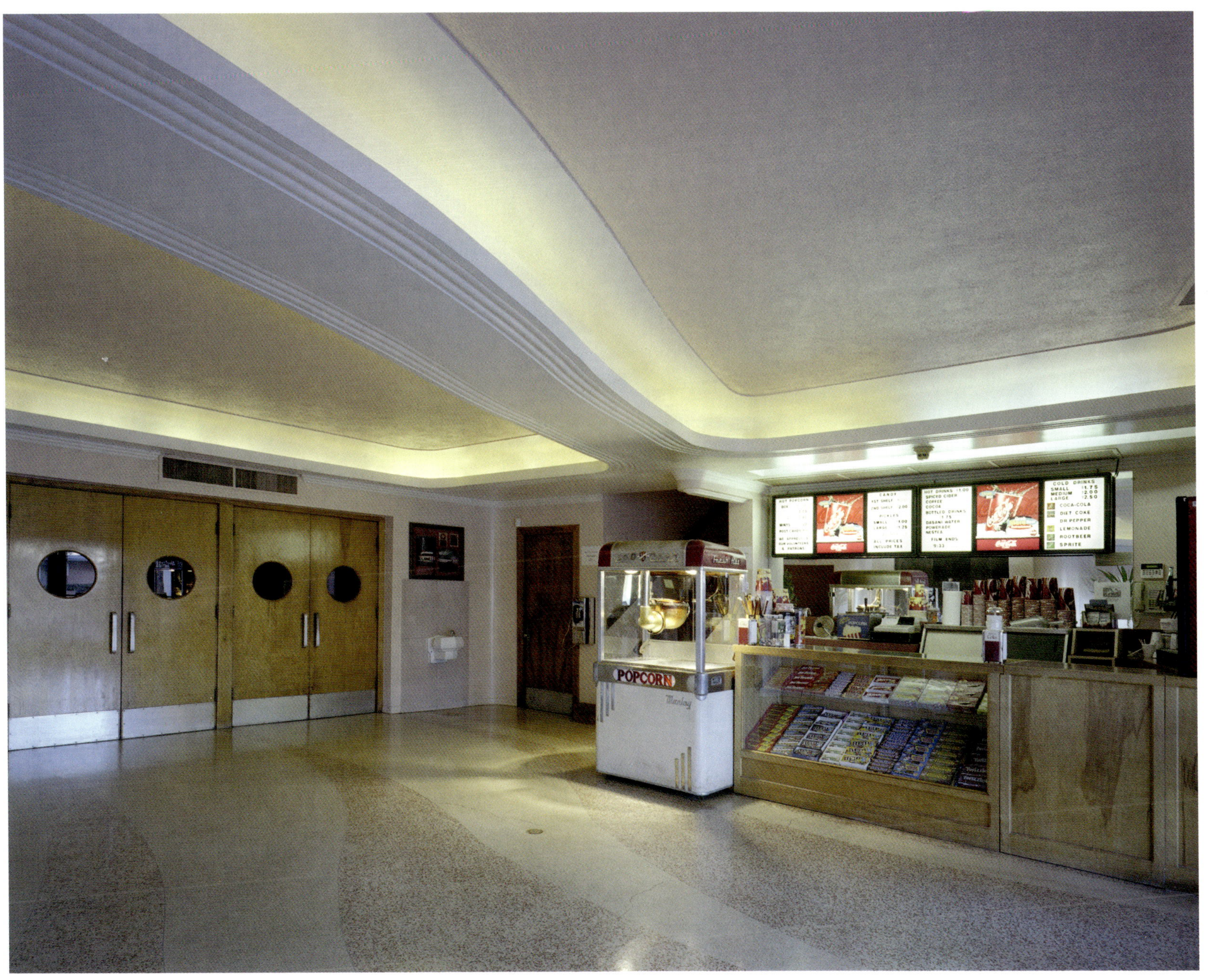

Foyer, Majestic Theatre, Eastland, Texas

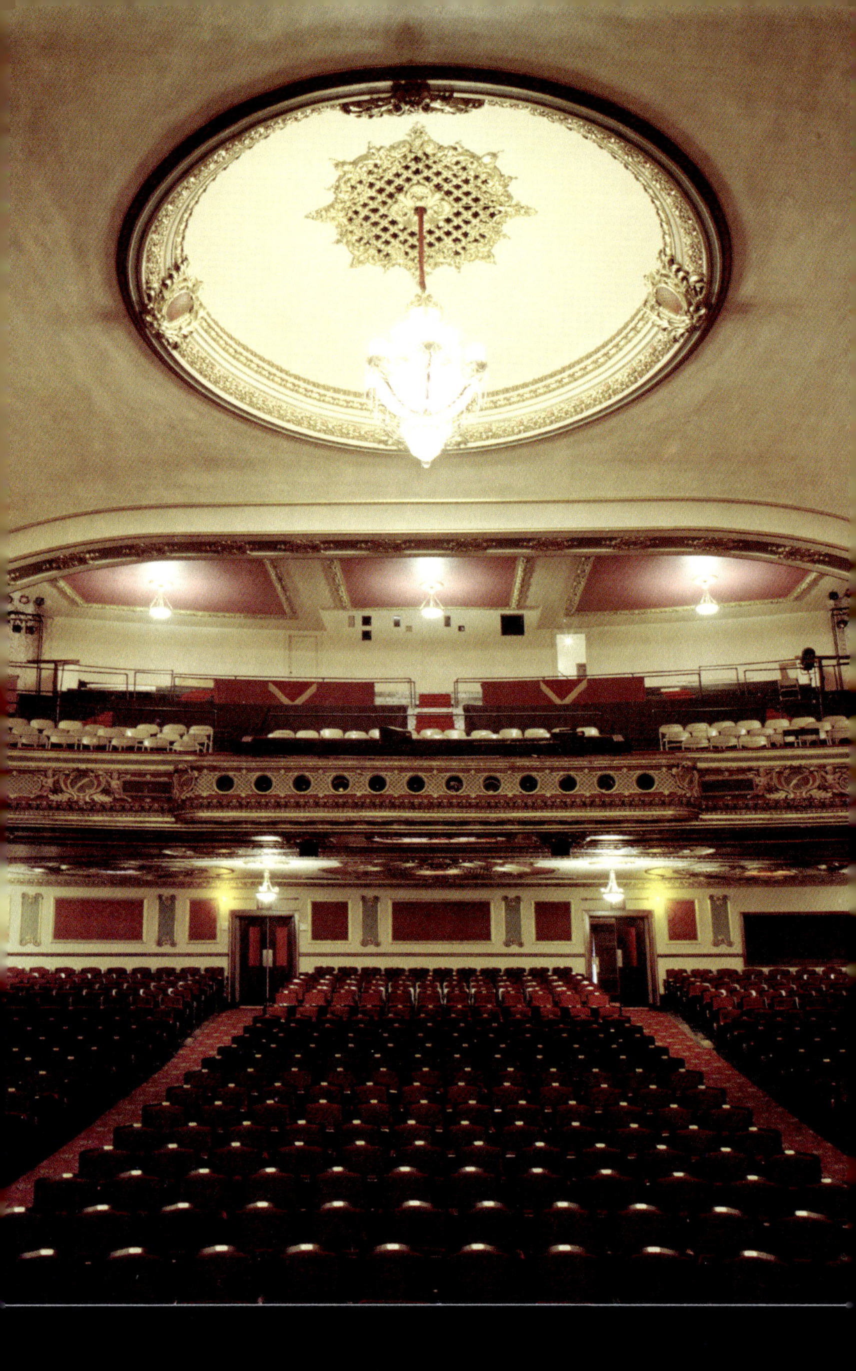

Interior, Coleman Theatre, Miami, Oklahoma

This is arguably the finest "small-town" theatre anywhere in the USA.

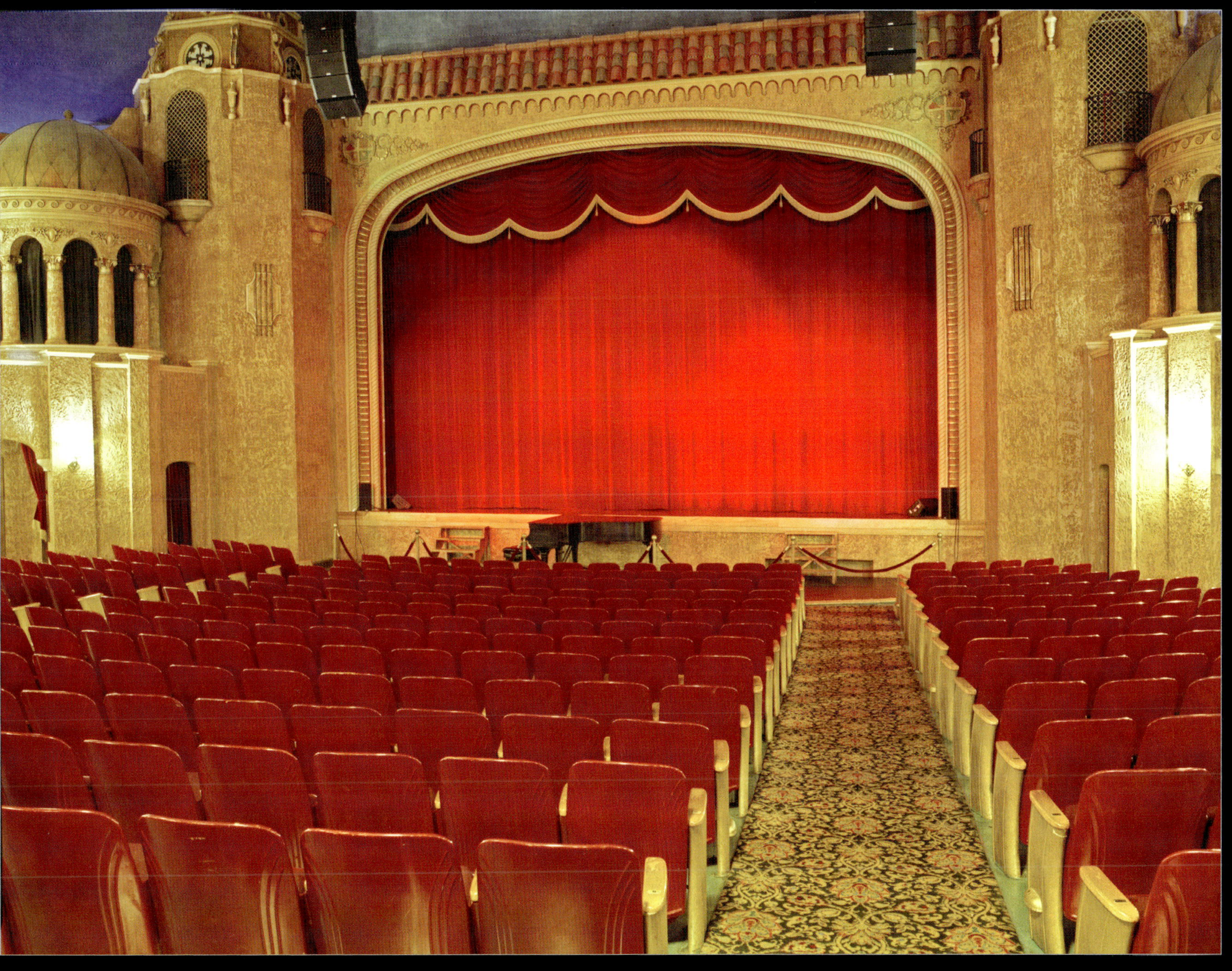

The historic Paramount Theatre, Abilene, Texas

One of many splendid movie theatres that now operates as a performing arts venue, the Paramount is on the National Register of Historic Places

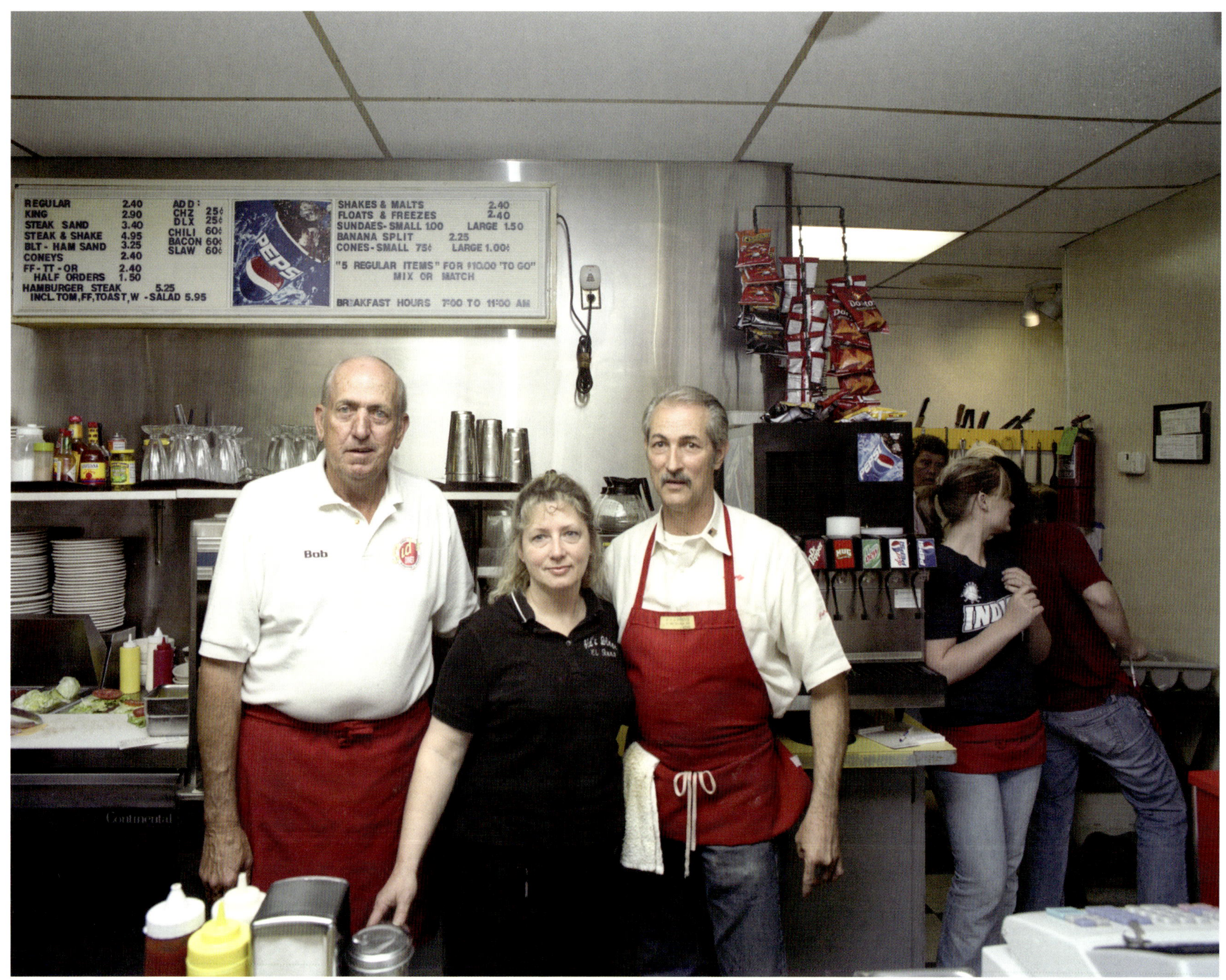

Bob, Carolyn and Marty, owners of the famous "Sid's Diner", El Reno, Oklahoma

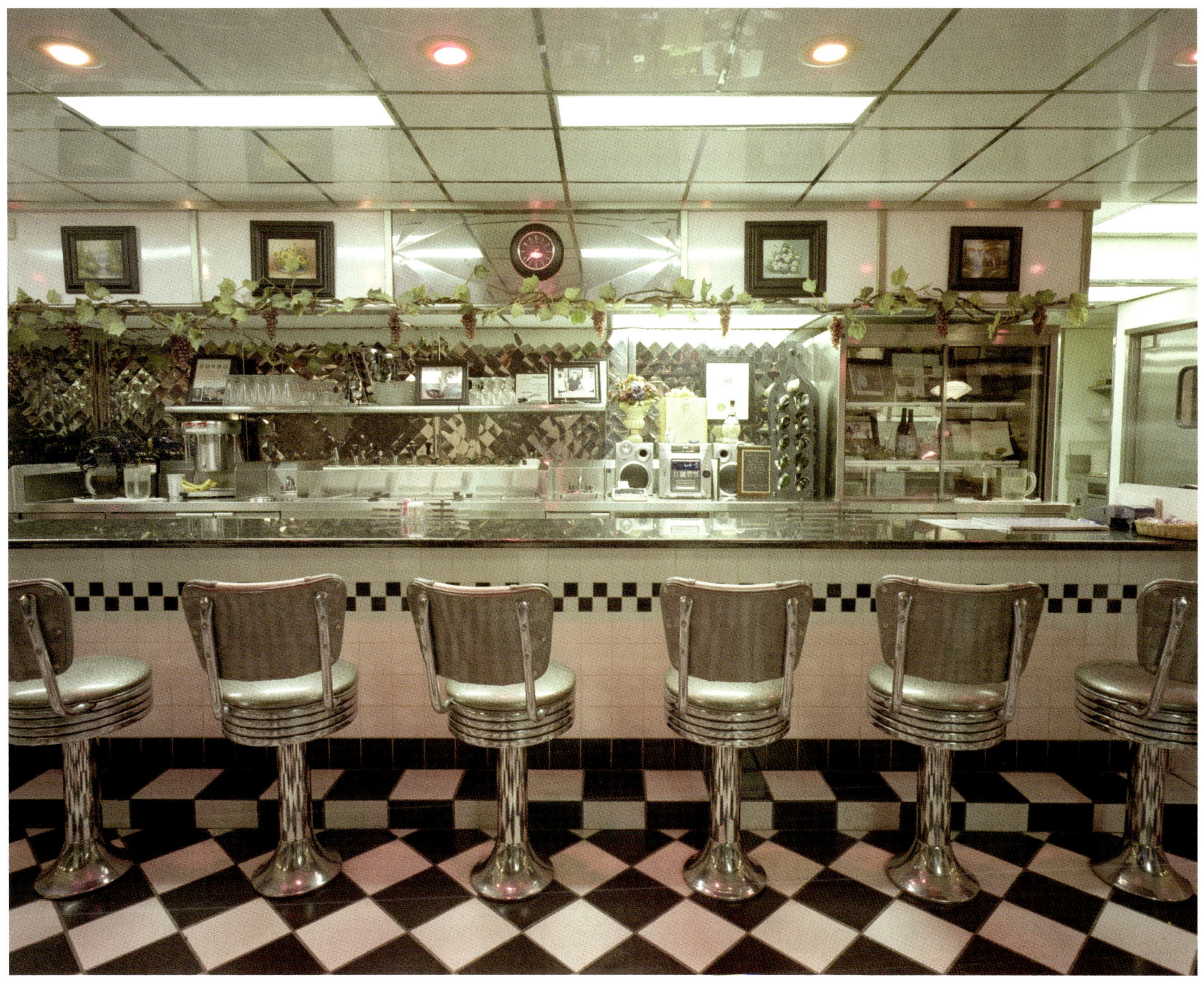

Interior, Valentino's Diner, Eastland, Texas

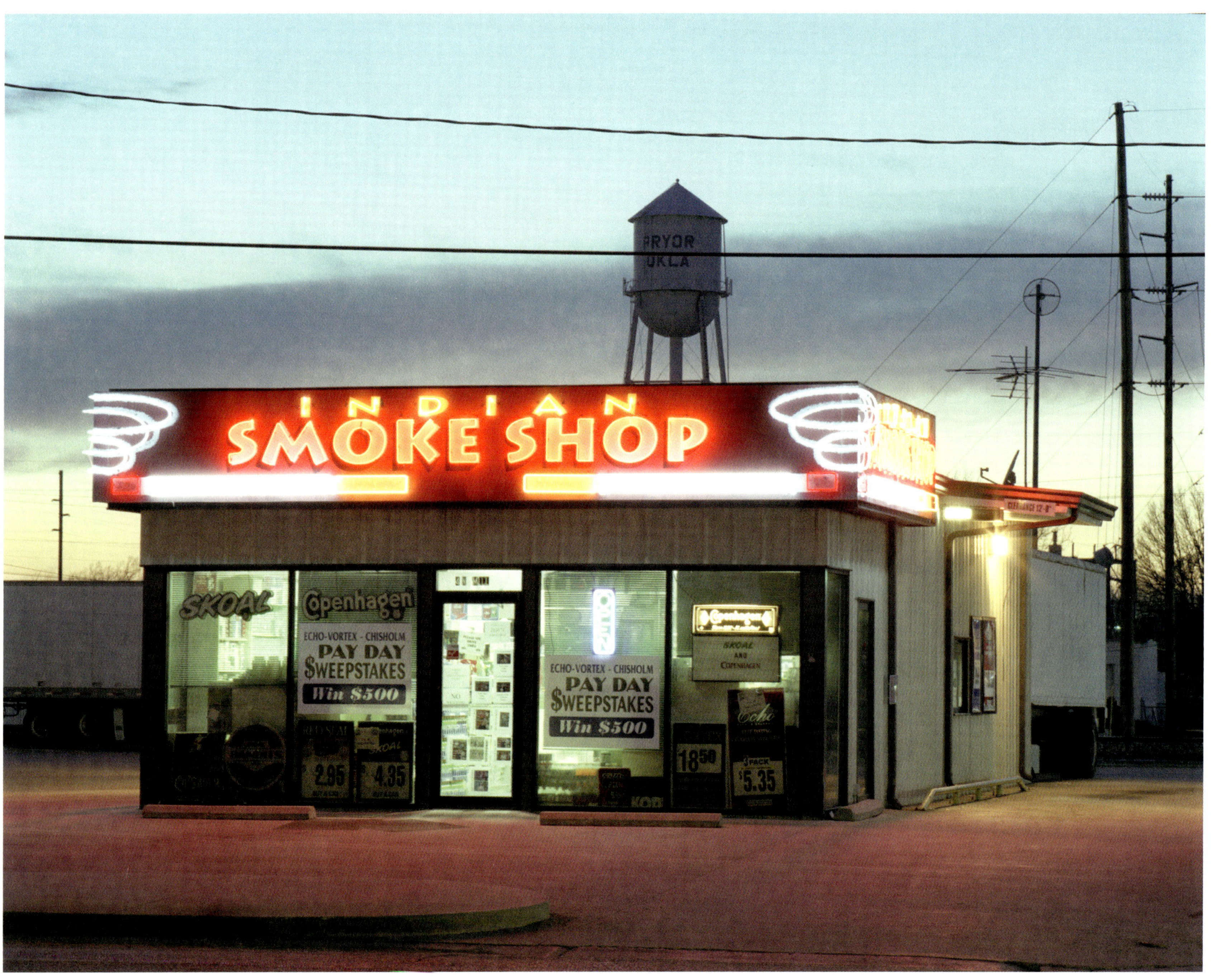

Indian Smokeshop, Pryor, Oklahoma

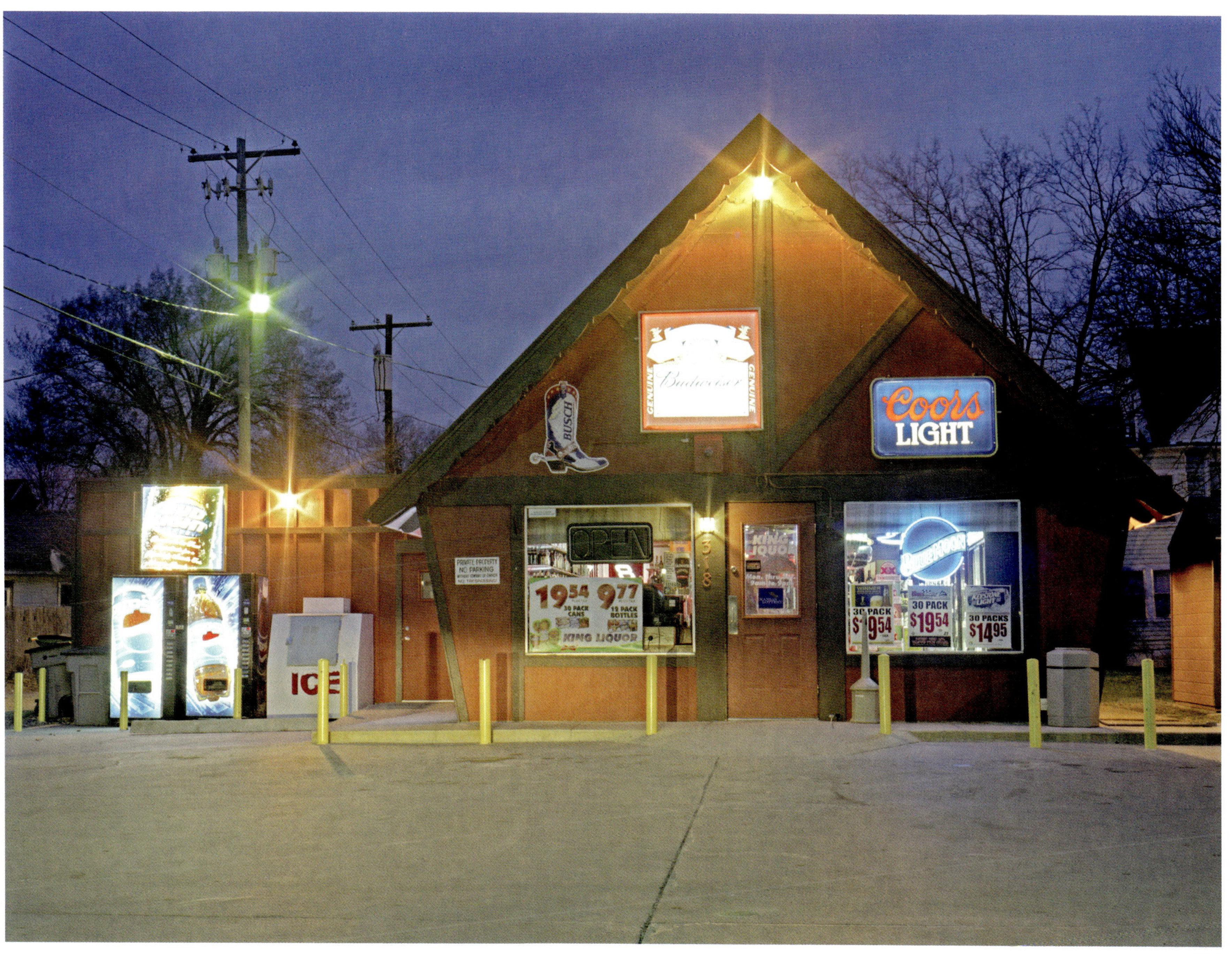

Liquor store, Emporia, Kansas

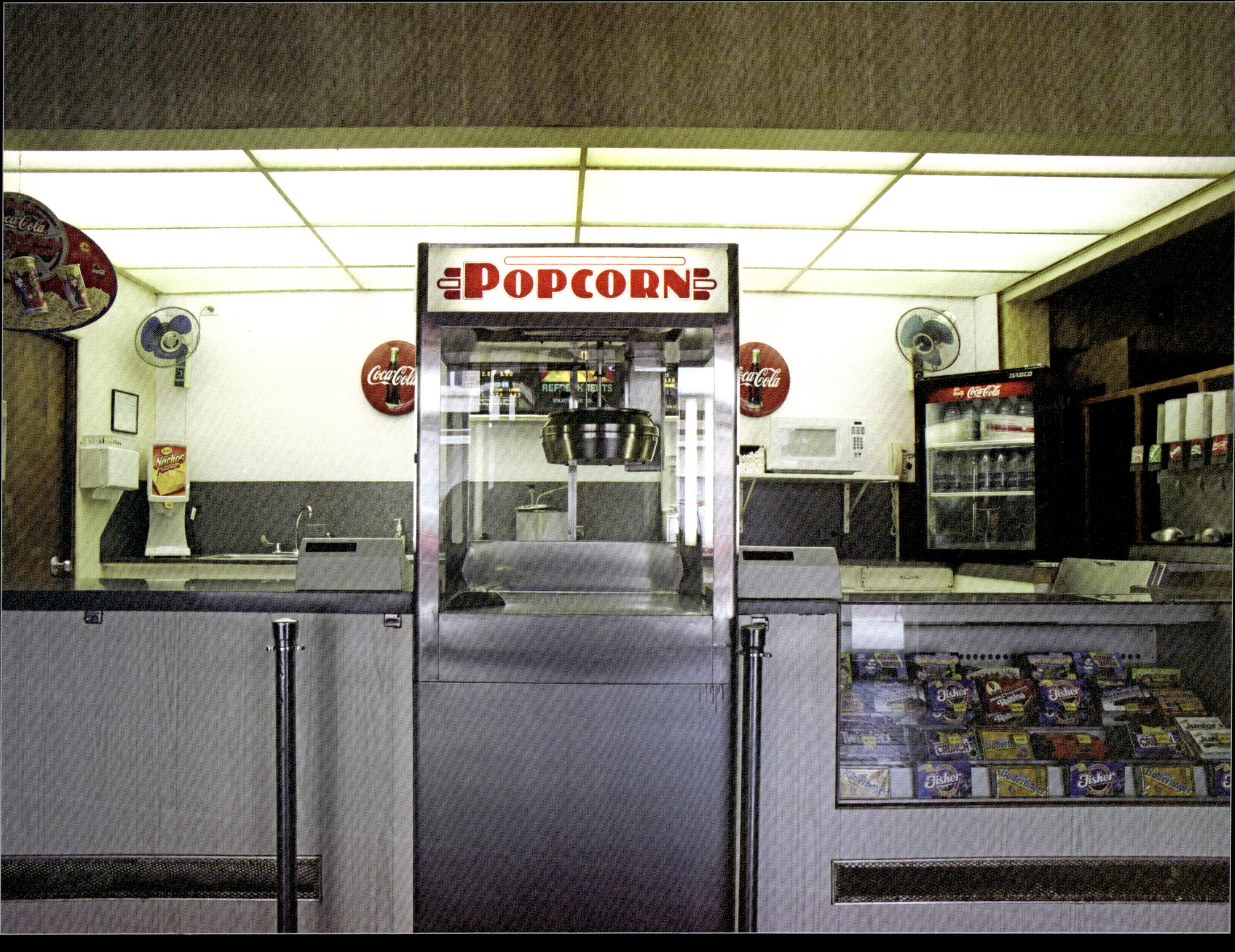

Popcorn, Hornbeck Theatre, Shawnee, Oklahoma

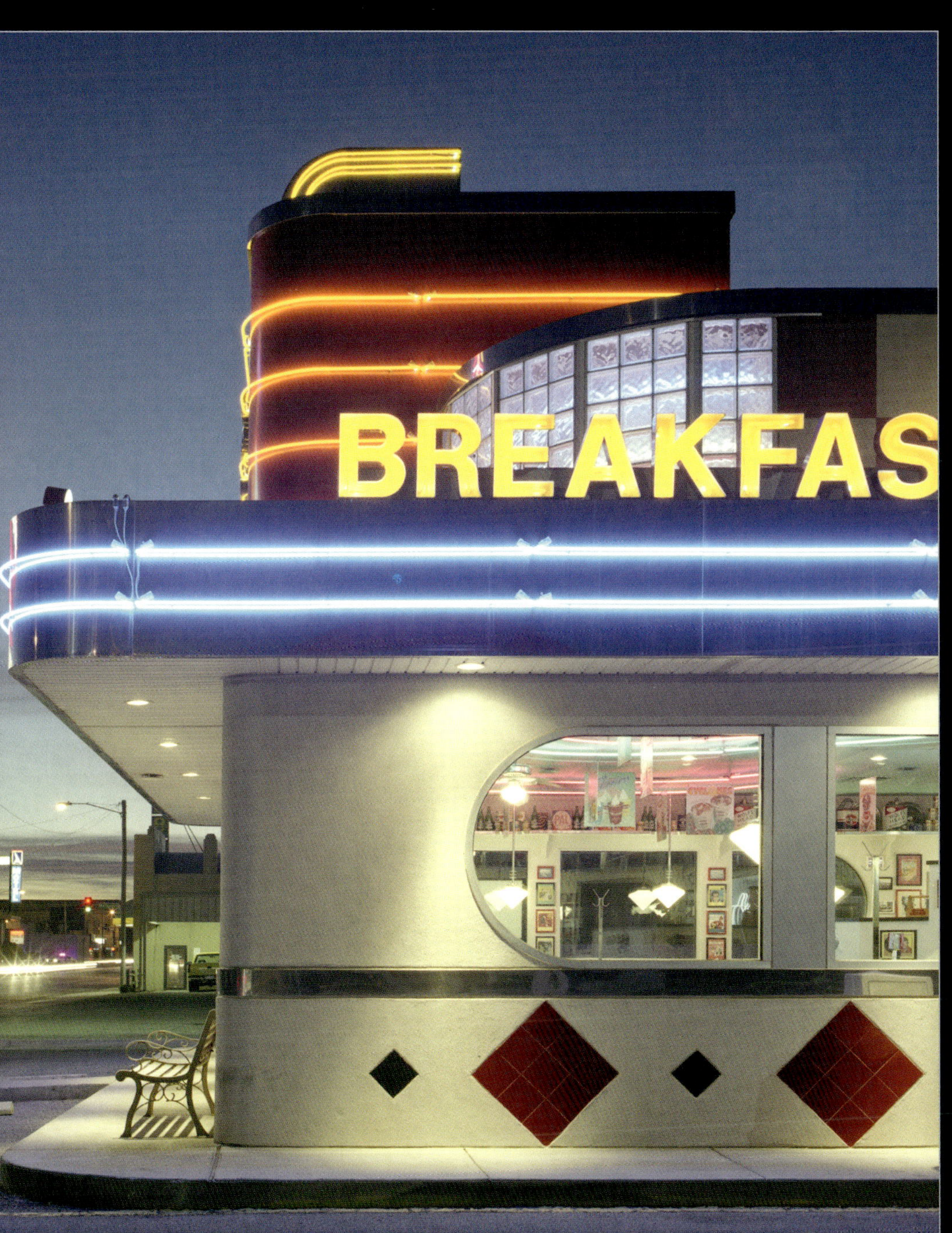
BREAKFAS

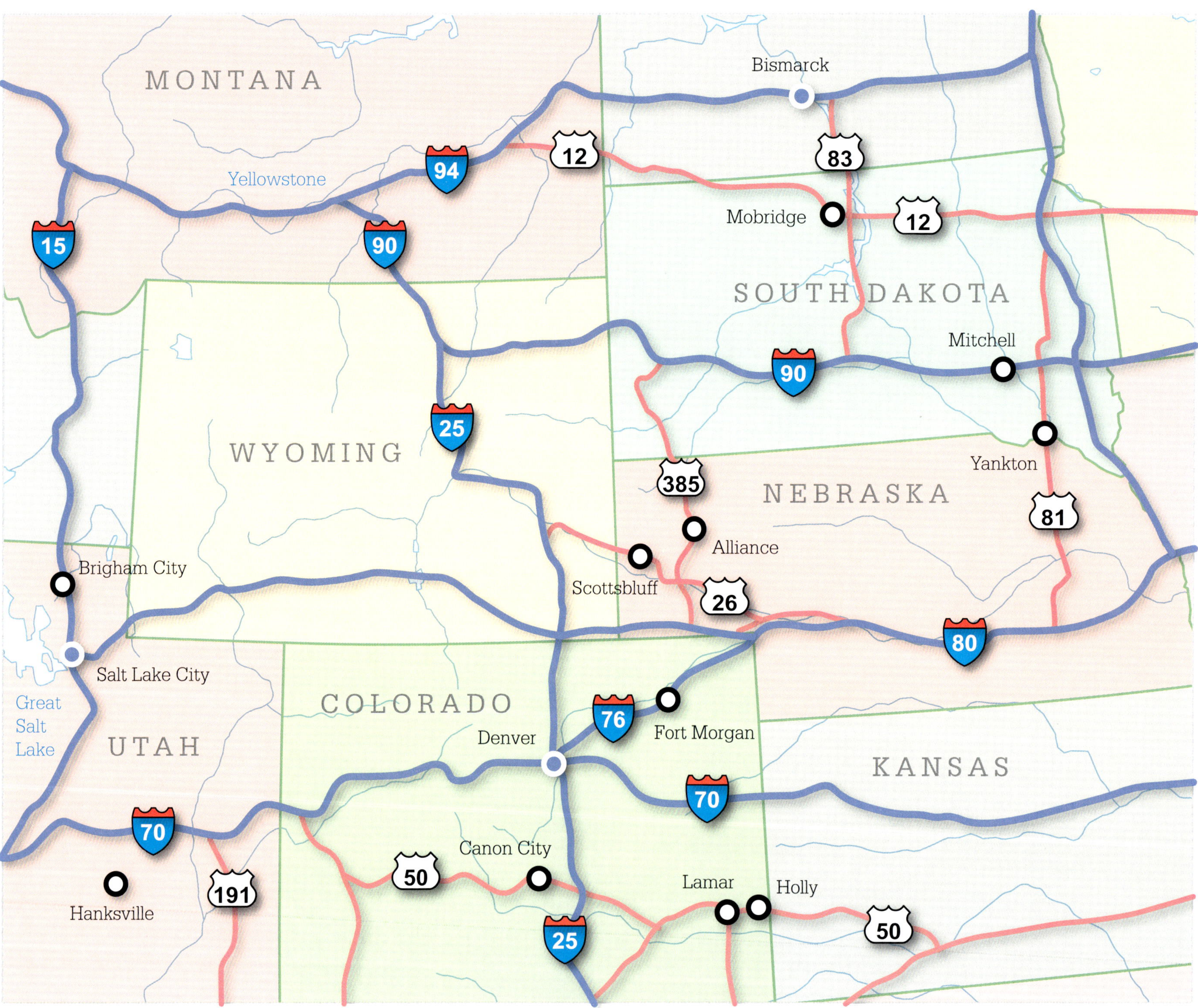

MONTANA
Yellowstone
Bismarck
SOUTH DAKOTA
Mobridge
Mitchell
Yankton
NEBRASKA
WYOMING
Alliance
Scottsbluff
Brigham City
Salt Lake City
Great Salt Lake
UTAH
COLORADO
Denver
Fort Morgan
KANSAS
Hanksville
Canon City
Lamar
Holly
94
12
83
12
15
90
90
25
385
81
26
80
76
70
70
191
50
25
50

Over the Rockies to the Prairies

Starting from Panguitch, Utah, the trip from Utah over the Rockies onto the prairies takes you through some of the most iconic landscapes America has to offer. Panguitch itself is a squeaky-clean Mormon town well placed to be a center for visiting several of the country's finest national parks, including Bryce Canyon, Zion National Park and the Escalante National Monument.

As you go north and east along Highway 12, the road steadily rises, passing the Dixie National Forest with its immense spread of ponderosa pine. From Teasdale the route weaves through the gigantic, spectacular rust-colored sandstone canyons of Capitol Reef National Park. Here you see towering spires of sculpted red rock, amazing arches and delicately eroded layers of shale. After you pass through Capitol Reef toward Hanksville the scenery changes, but the dry barren landscape, with its dramatic isolated buttes, remains equally impressive.

Northward past the Goblin Valley State Park, the red-stone theme continues, although everything appears slightly diminished in scale. The soaring cathedrals of rock give way to smaller "goblins", spires and precariously balanced boulders. By way of contrast, the middle and northern parts of Utah offer a more fertile environment able to support countless farms and ranches. This area was once rich in coal, and so place-names like Carbon and Coalville abound. By contrast, Brigham City was named to honor one of the founding fathers of the Mormon Church. Set within an arable landscape, it is situated very close to a large area of freshwater marshes created by the meeting of the Bear River and the Great Salt Lake.

Of the limited number of routes that cross the Rockies into Colorado, US 50 – romantically dubbed the Loneliest Road – is undoubtedly among the most appealing. Leaving Green River, it passes close to some of America's most stunning scenery, including Canyonlands and Arches National Park, before rising steeply to over 10,000 feet. Here there are traces of snow visible even in the height of summer. As the road descends into Colorado, the landscape is characterized at first by alpine meadows and deep river canyons, but these in turn give way to flat agricultural prairies once the mountains are behind you.

Colorado is a state of contrasts. The west is dominated by the Rockies and the fashionable ski resorts. Further east the land becomes flatter, ideal for the numerous ranches and farms. Just east of the Rockies, Cañon City is still over a mile high, with peaks of 14,000 feet visible on the horizon. It is a classic Wild West town with its saloons, gun-shops and saddleries. It is not until you have passed the heavily industrialized town of Pueblo that you arrive in the prairies proper.

The contrast with Utah and the Rockies could not be starker. This is a land of vast open plains and, despite the agriculture, the absence of trees creates an almost barren appearance. Other vehicles are a rare sight, although occasionally you see in the distance a plume of dust as some SUV hares across the prairie to join Highway 71 several miles ahead. Typically the roads are long and straight, passing through almost forgotten communities such as Limon, Last Chance and Woodrow.

Highway 71 now continues north over the border into Nebraska, going past Scottsbluff and up to the town of Chadron, where you can pick up US 20, otherwise known as the Oregon Trail. Bearing in mind that Nebraska is often labeled one of America's most boring states, the scenery here is surprisingly varied, starting with the vast and undulating Sandhills in the west. The middle of the state is sparsely

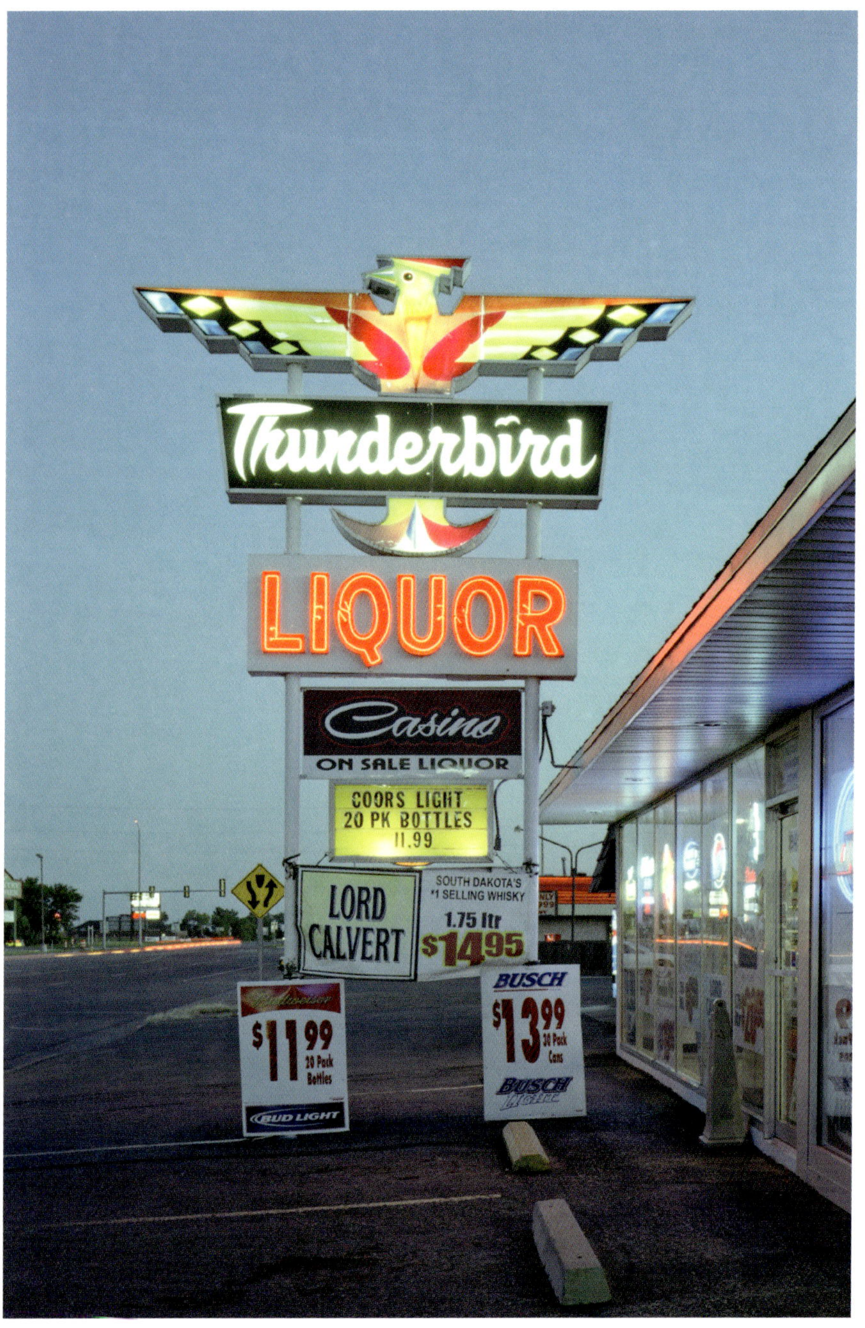

Thunderbird Liquor, Mobridge, South Dakota

populated and the towns few and far between; the landscape is rural in nature and has a gentle rolling character that is pleasant to drive across. To the east of this area is the town of O'Neill, regarded as the Irish capital of Nebraska, a fact reinforced by a large green painted shamrock at the main intersection through town.

As you head north toward the Dakotas, just before you arrive in Yankton, you cross the Missouri, an experience that offers a reminder of just how large and impressive so many of the American rivers are. In South Dakota you find large tracts of farmland with occasional hills and ravines. The journey to Mitchell takes you through a myriad of small towns, each with their gas station, food store and agricultural-machine depot. At Pierre you can pick up US Highway 83, which, in this thinly populated pocket of America, is perhaps appropriately nicknamed the Road to Nowhere. The area to the east of Pierre is known as the Great Lakes, but if instead you venture north toward Selby, passing small towns such as Agar and Onide, you encounter farmlands where Norman Rockwell-style sleepy hamlets and picturesque grain elevators abound. Between Pierre and Selby, a particularly large Cathedral on the Prairie looms over the small community of Hoven.

At the northern end of the state, US 83 passes close to the large Standing Rock Indian Reservation (home to 4,500 Sioux) and also the town of Mobridge, another point where the imposing Missouri River can be crossed. Due to its location on Lake Oahe, Mobridge is a magnet for fishermen, while the surrounding prairies attract hunters of pheasant, grouse and deer. This landscape of endless cultivated fields of green and gold, punctuated only by silos, grain elevators and the occasional abandoned homestead, typifies the quiet charm of so much of the American heartlands.

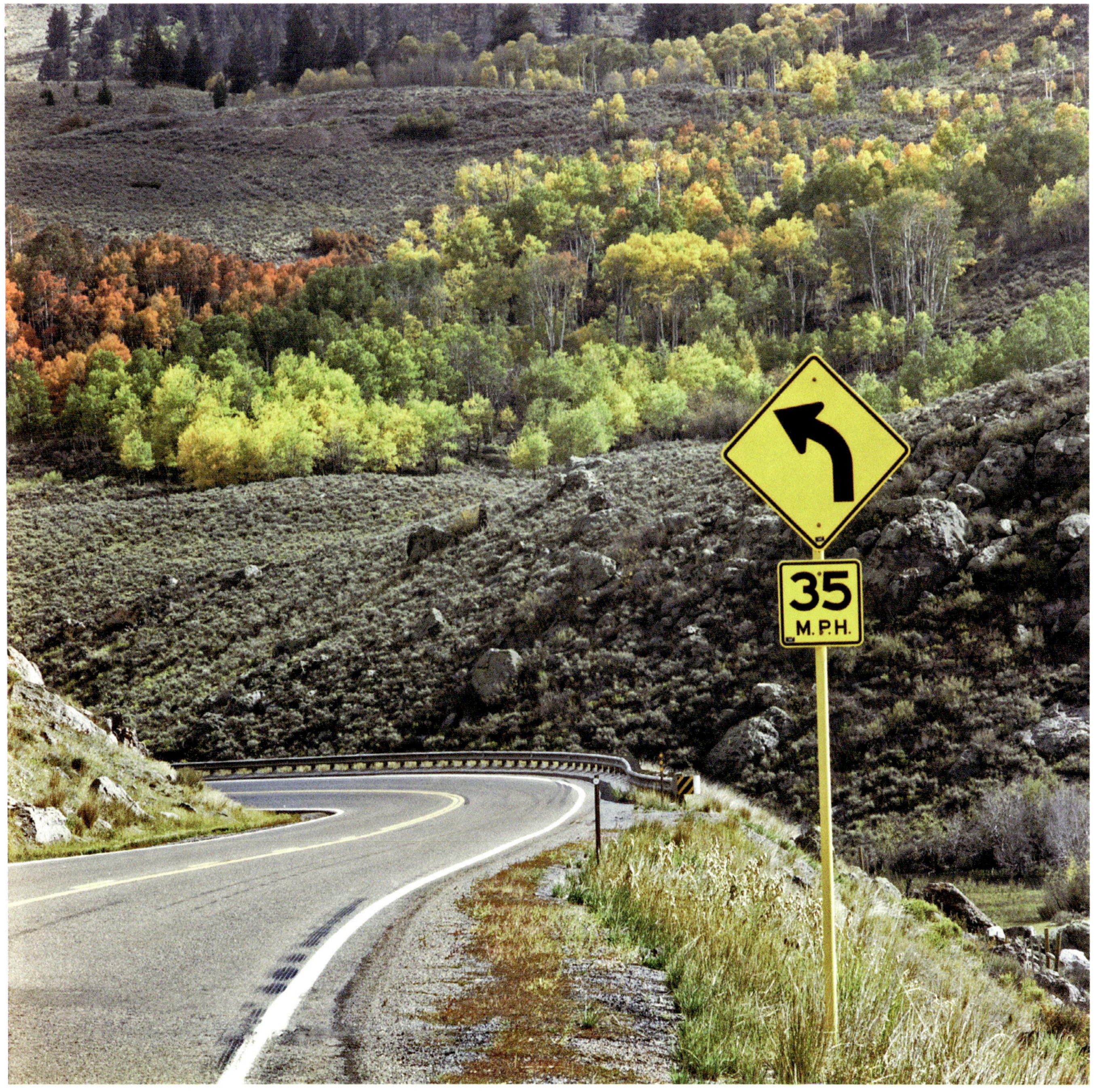

On US Hwy 50, romantically dubbed "The Loneliest Road", on the eastern slopes of the Rockies, Colorado

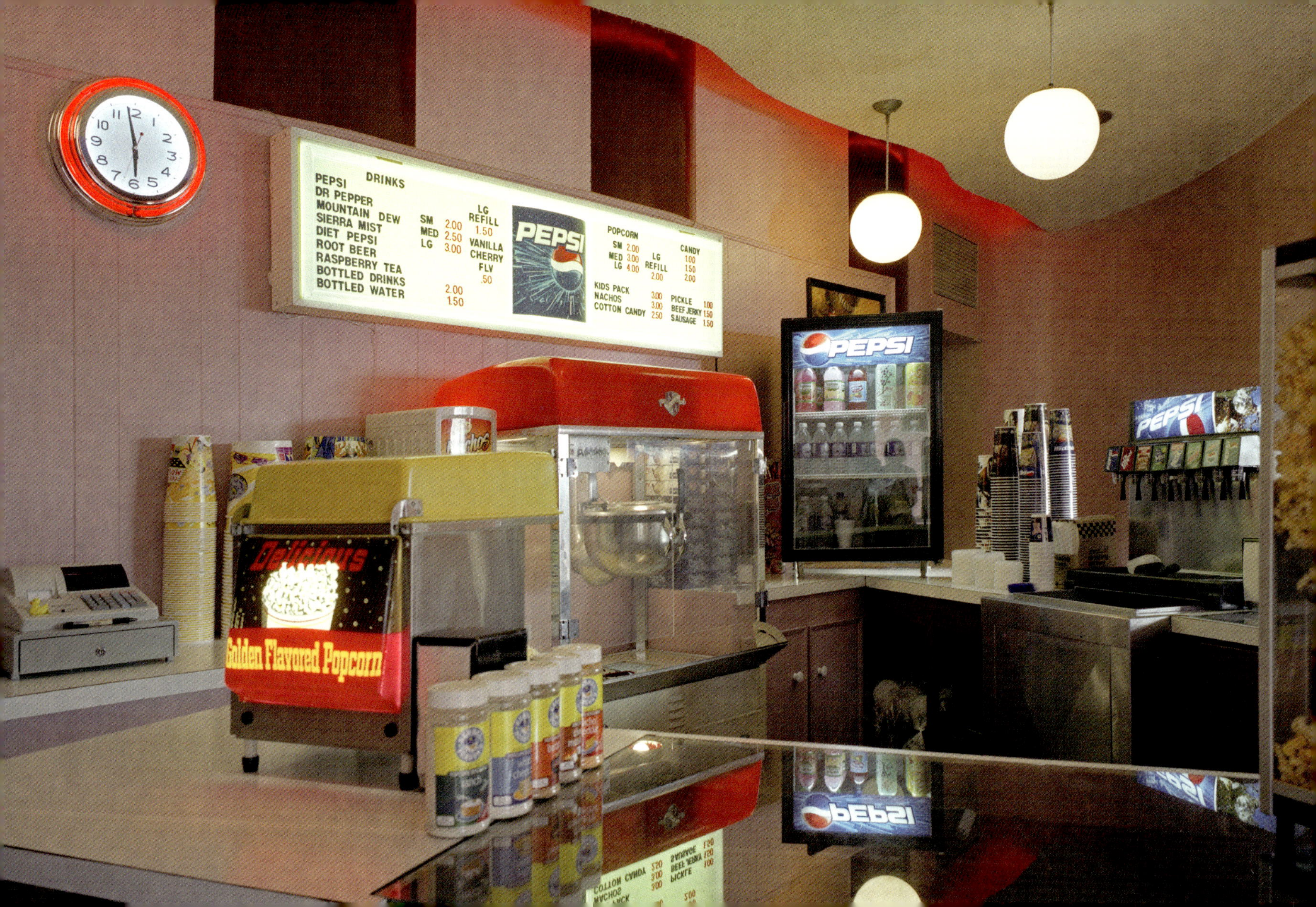

DRINKS
PEPSI
DR PEPPER
MOUNTAIN DEW
SIERRA MIST   SM   2.00
DIET PEPSI   MED   2.50
ROOT BEER   LG   3.00
RASPBERRY TEA
BOTTLED DRINKS
BOTTLED WATER   2.00
                        1.50
LG REFILL 1.50
VANILLA
CHERRY
FLV .50
PEPSI
POPCORN
SM 2.00
MED 3.00   LG
LG 4.00   REFILL
              2.00
CANDY
1.00
1.50
2.00
KIDS PACK
NACHOS   3.00
COTTON CANDY   3.00   PICKLE   1.00
               2.50   BEEF JERKY   1.50
                       SAUSAGE   1.50
Delicious
Golden Flavored Popcorn

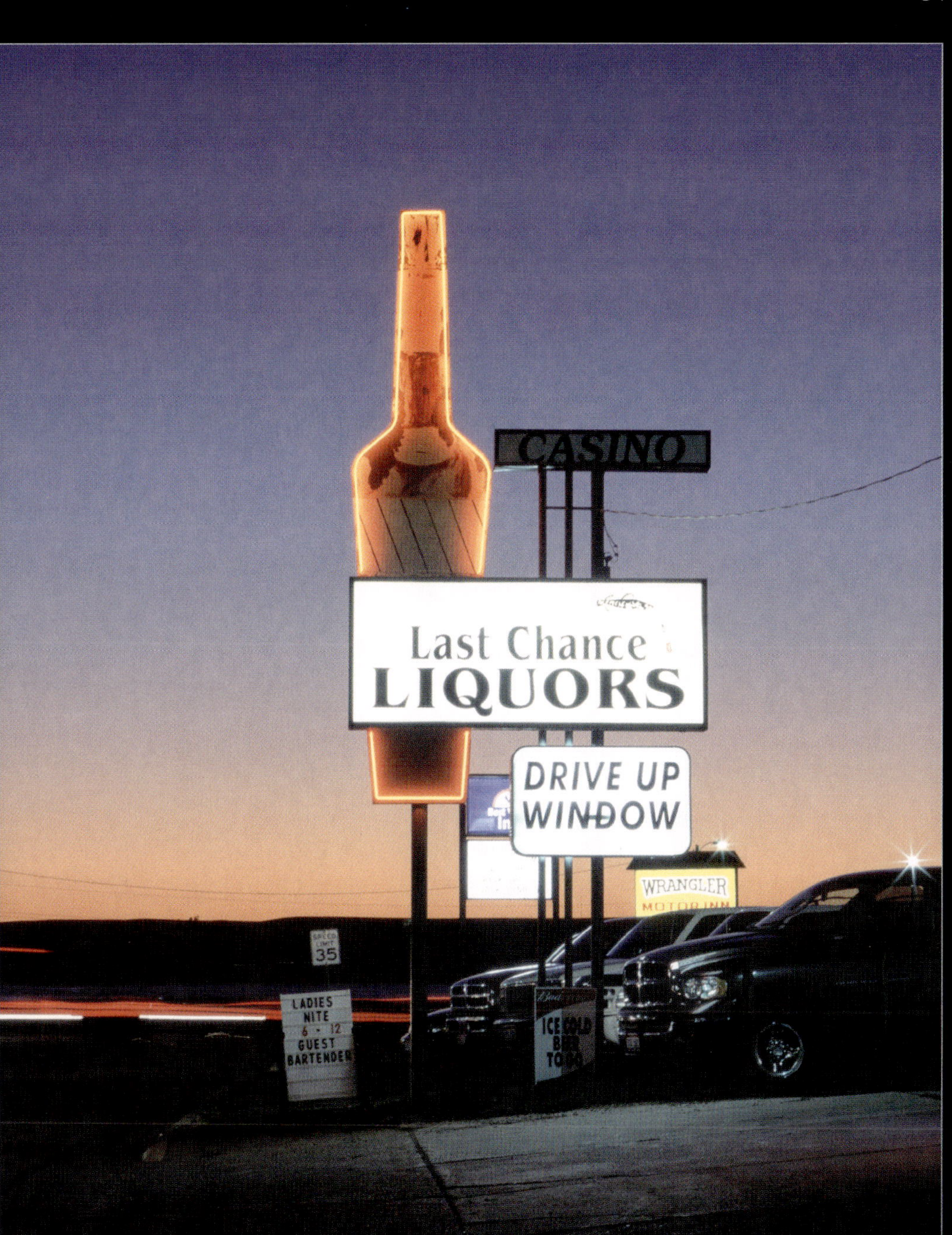
CASINO
Last Chance
LIQUORS
DRIVE UP
WINDOW
WRANGLER
MOTOR INN
SPEED
LIMIT
35
LADIES
NITE
6 - 12
GUEST
BARTENDER
ICE COLD
BEER
TO GO

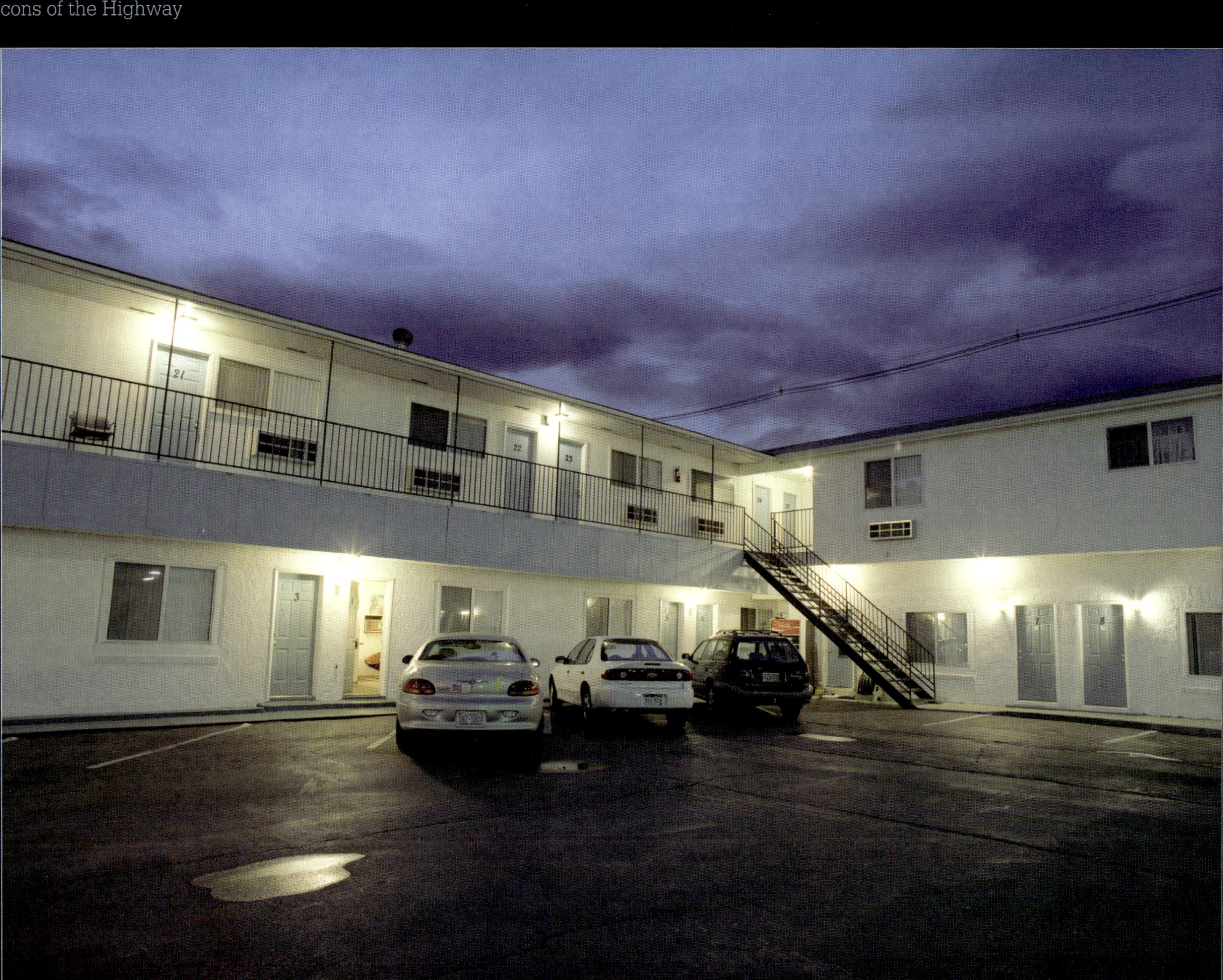

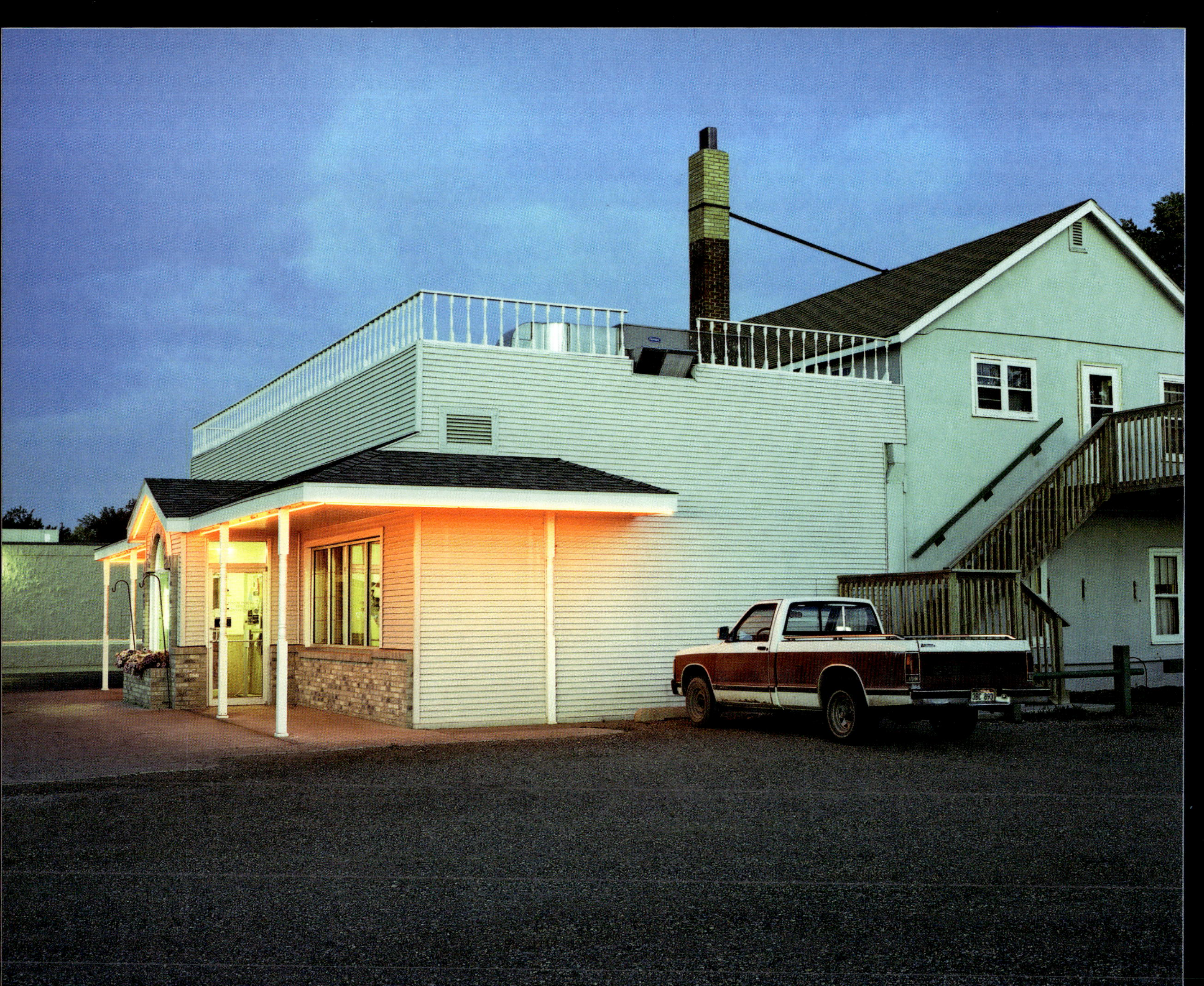

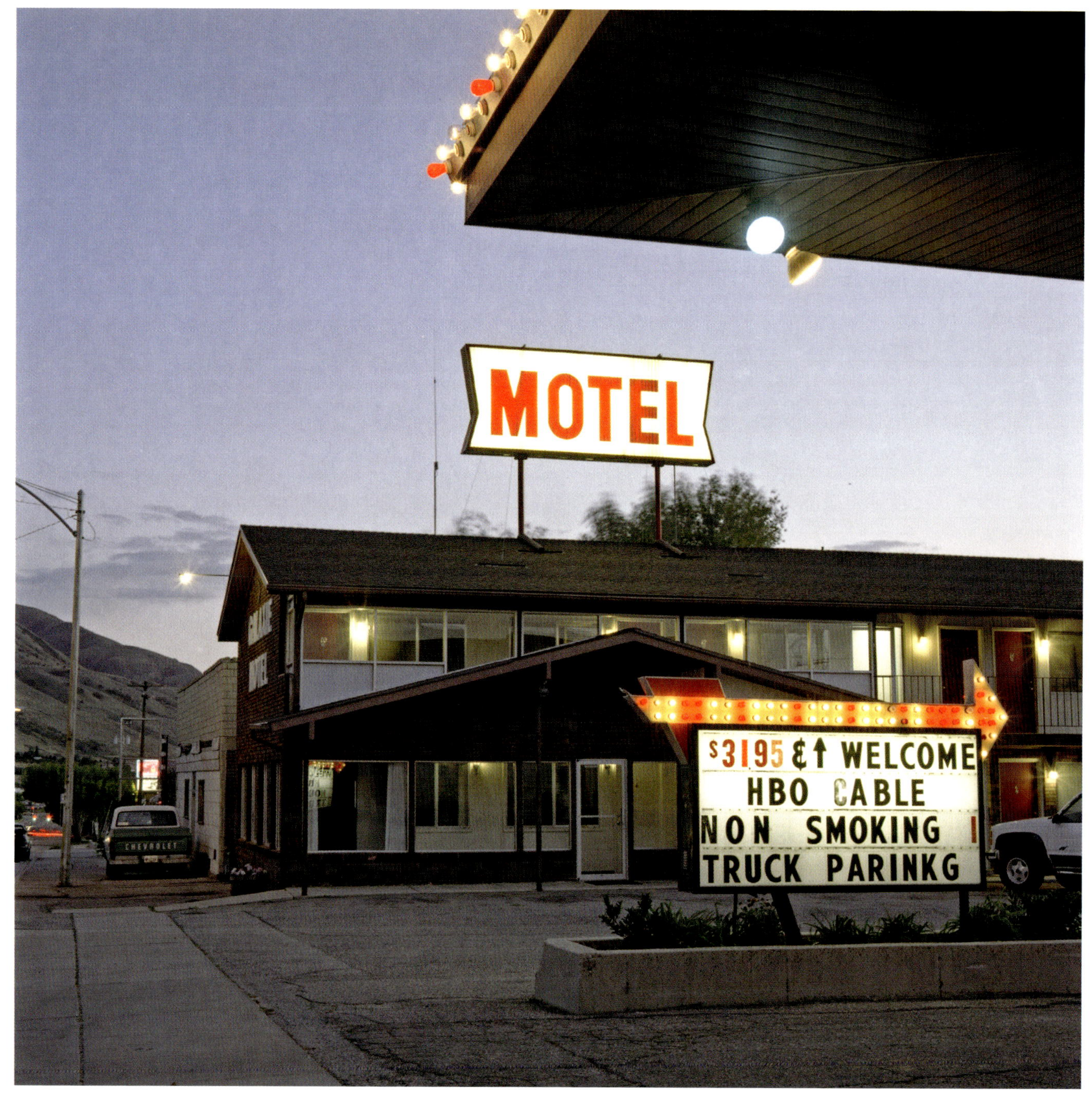

Motel, Brigham City, Utah

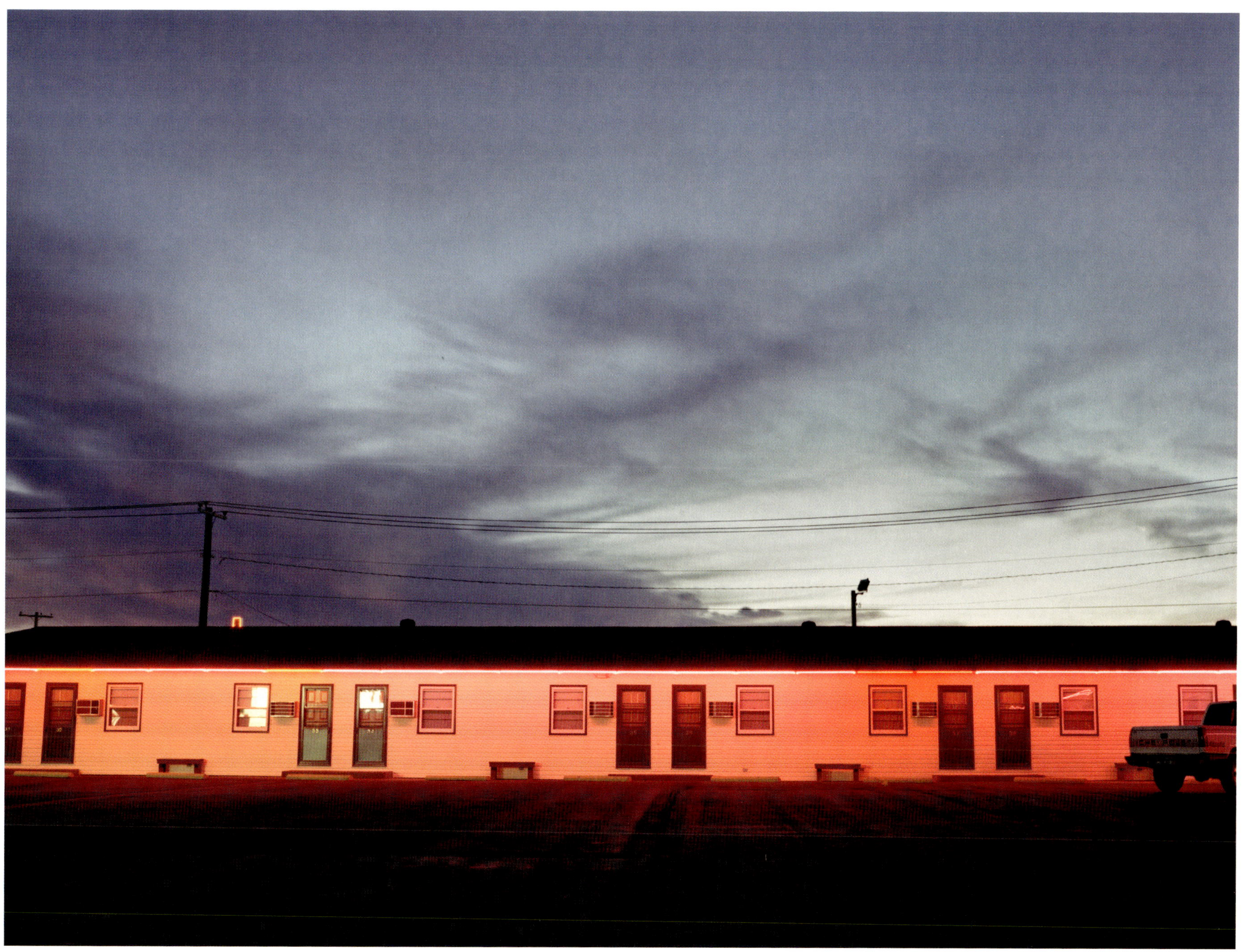

The Mo-Rest Motel, Mobridge, South Dakota

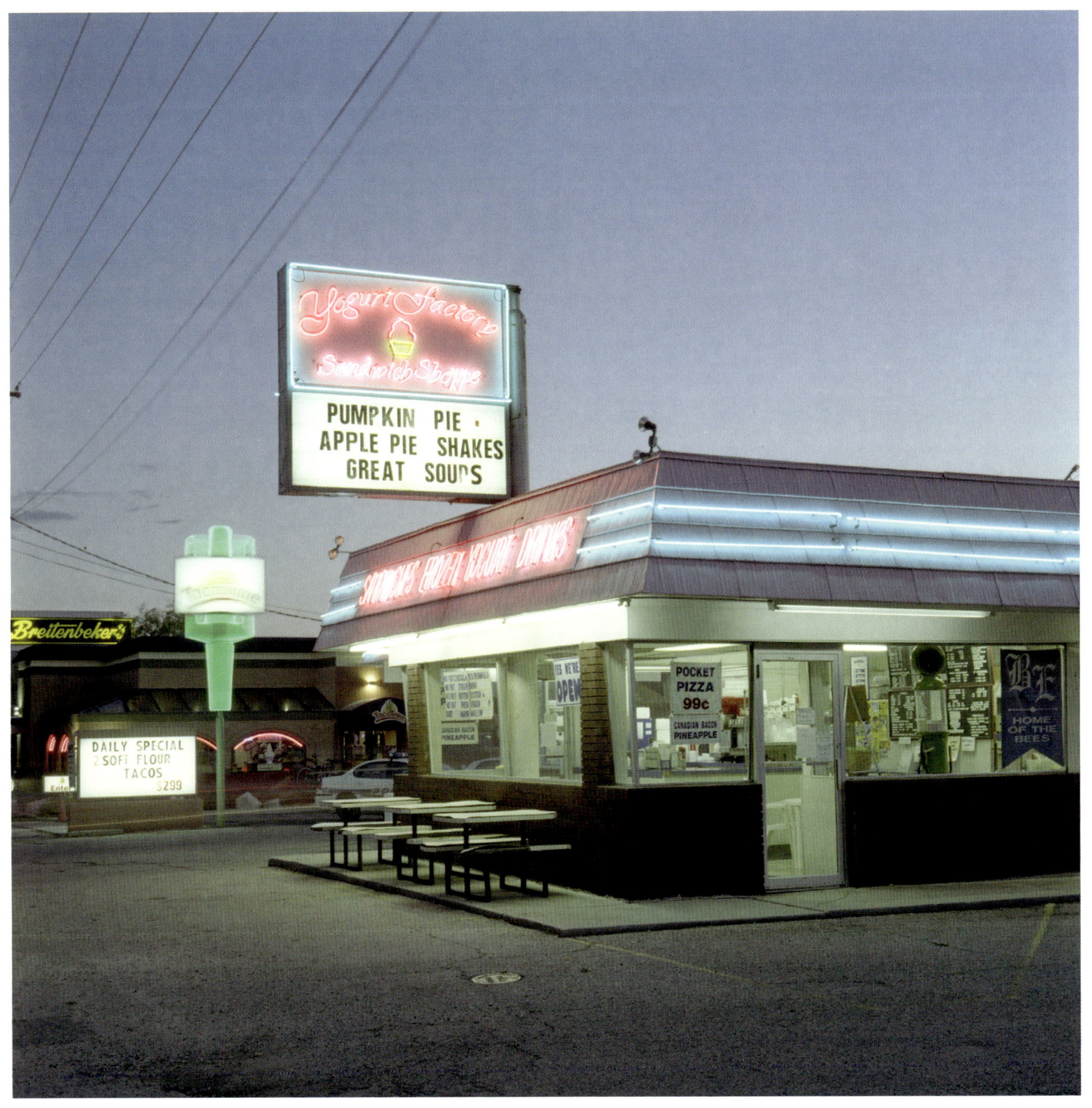

"Yogurt Factory", Brigham City, Utah

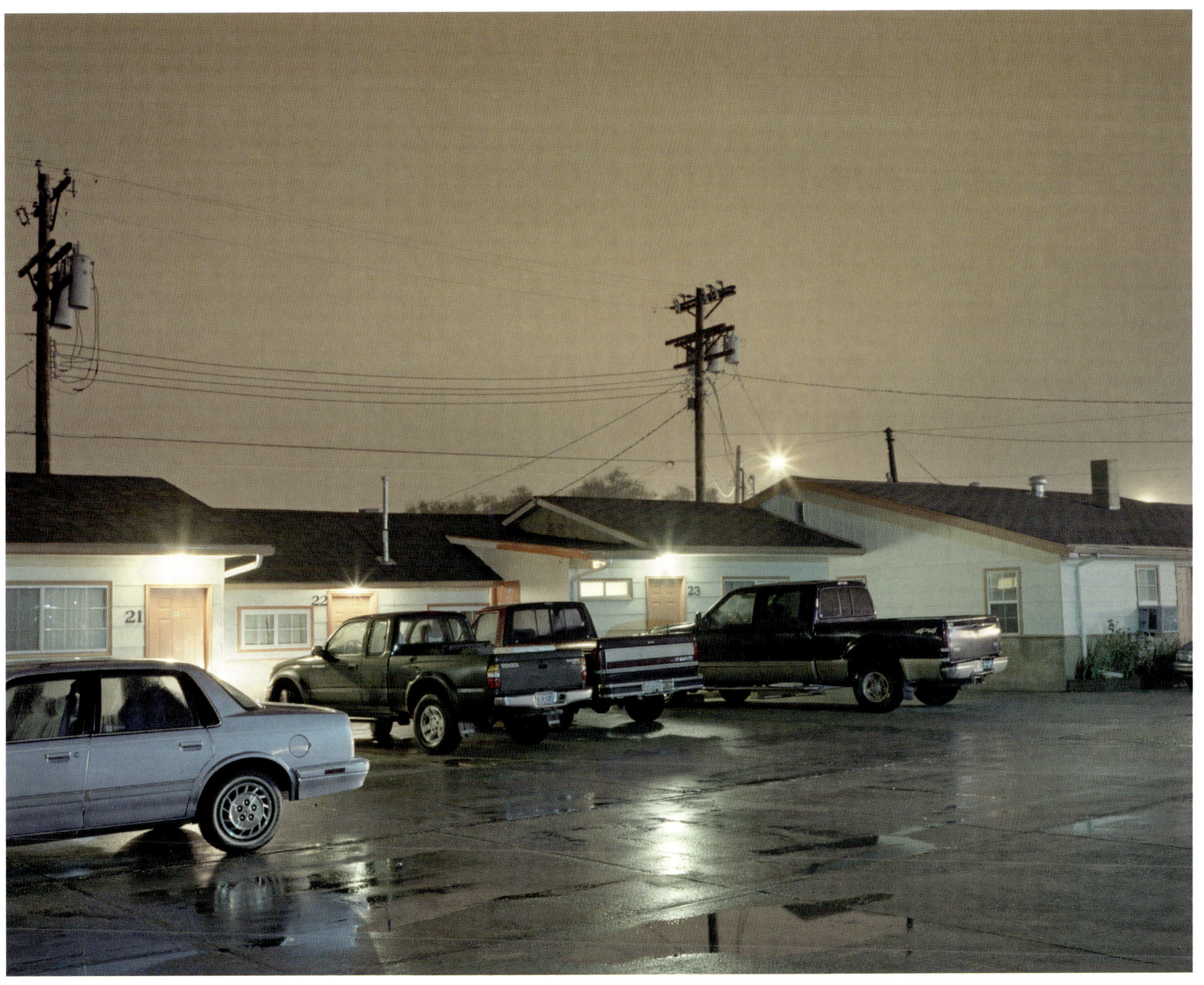

Motel parking lot, Alliance, Nebraska

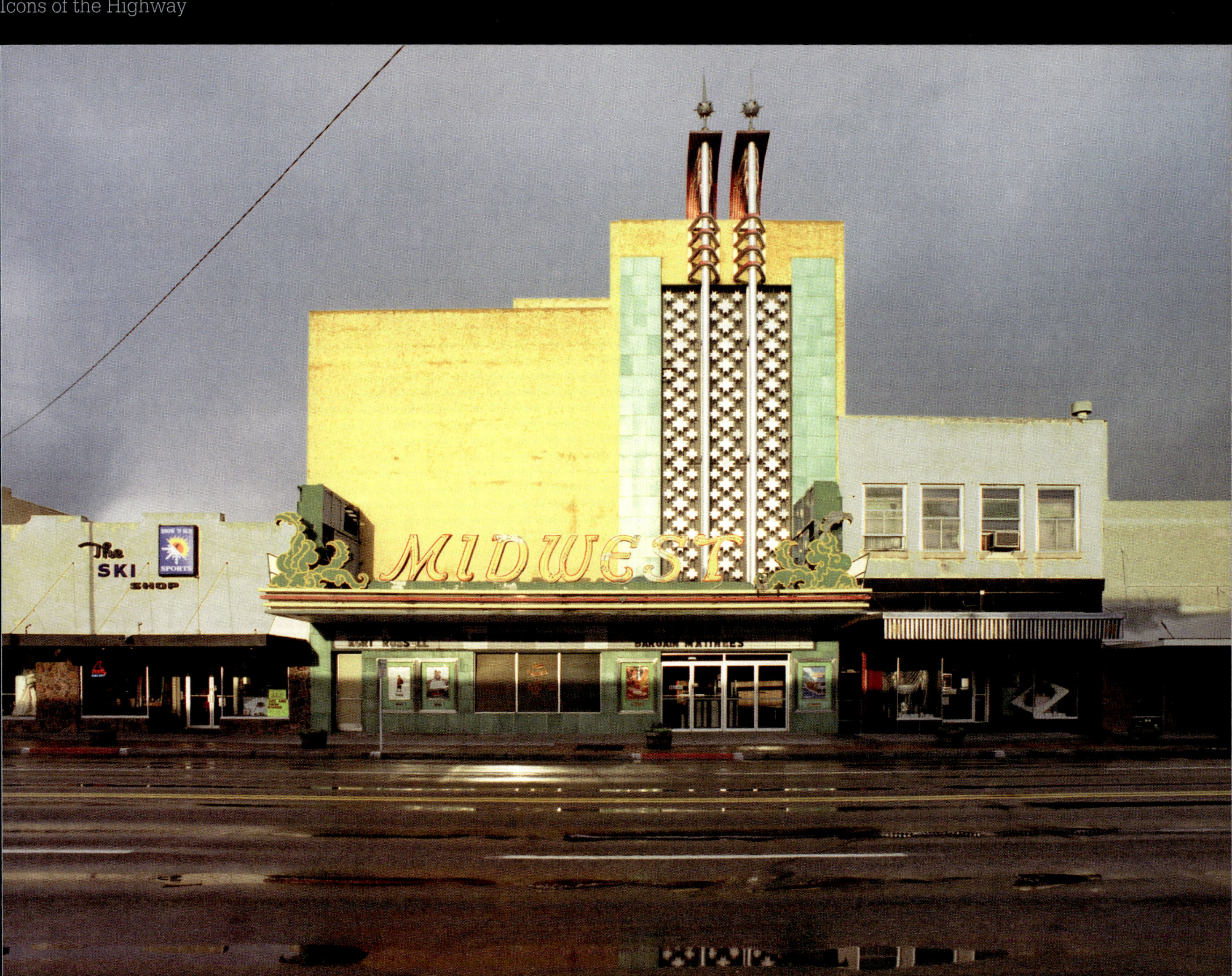
The SKI
SNOW 'N SKI
SPORTS
SHOP
MIDWEST
BARGAIN MATINEES

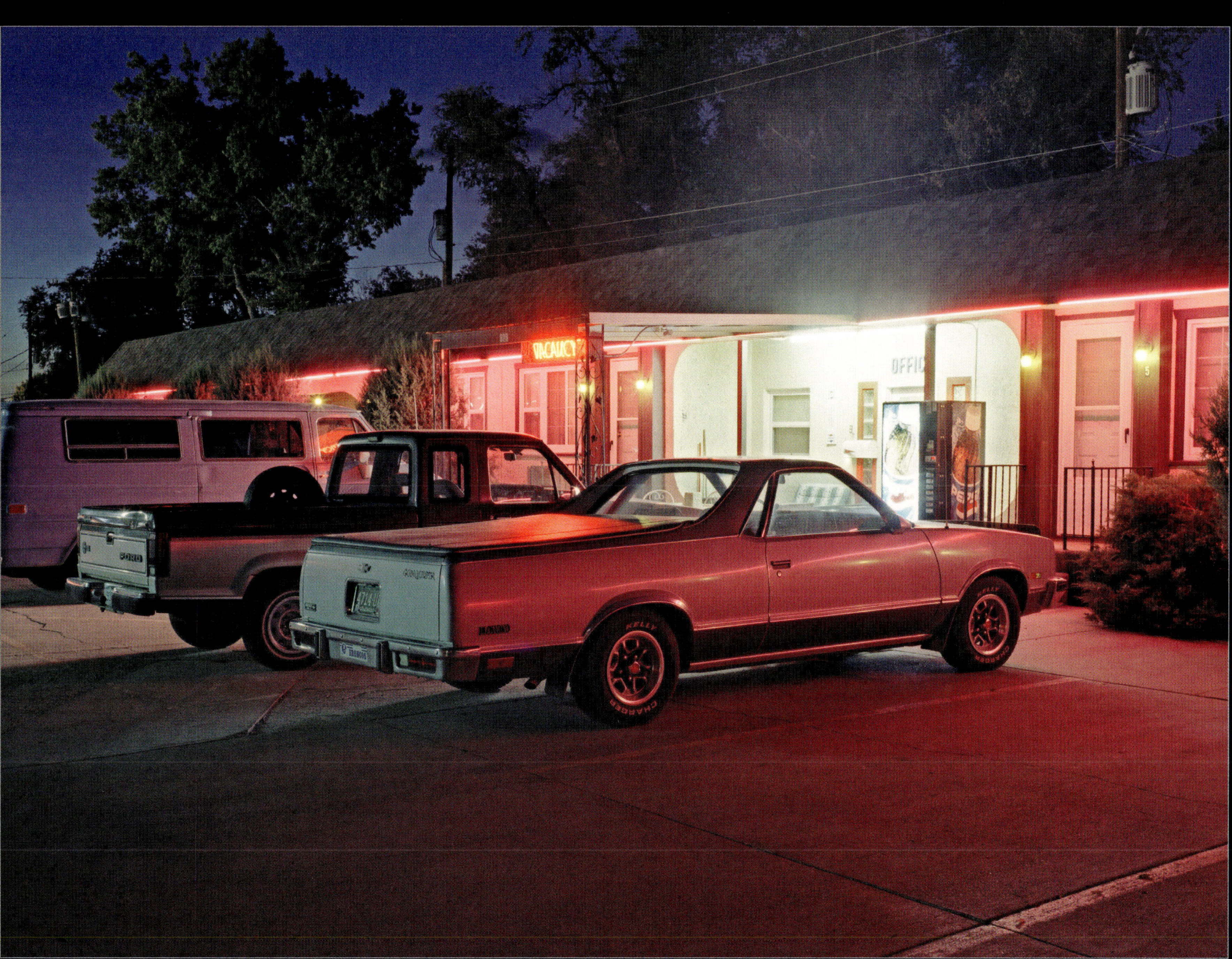

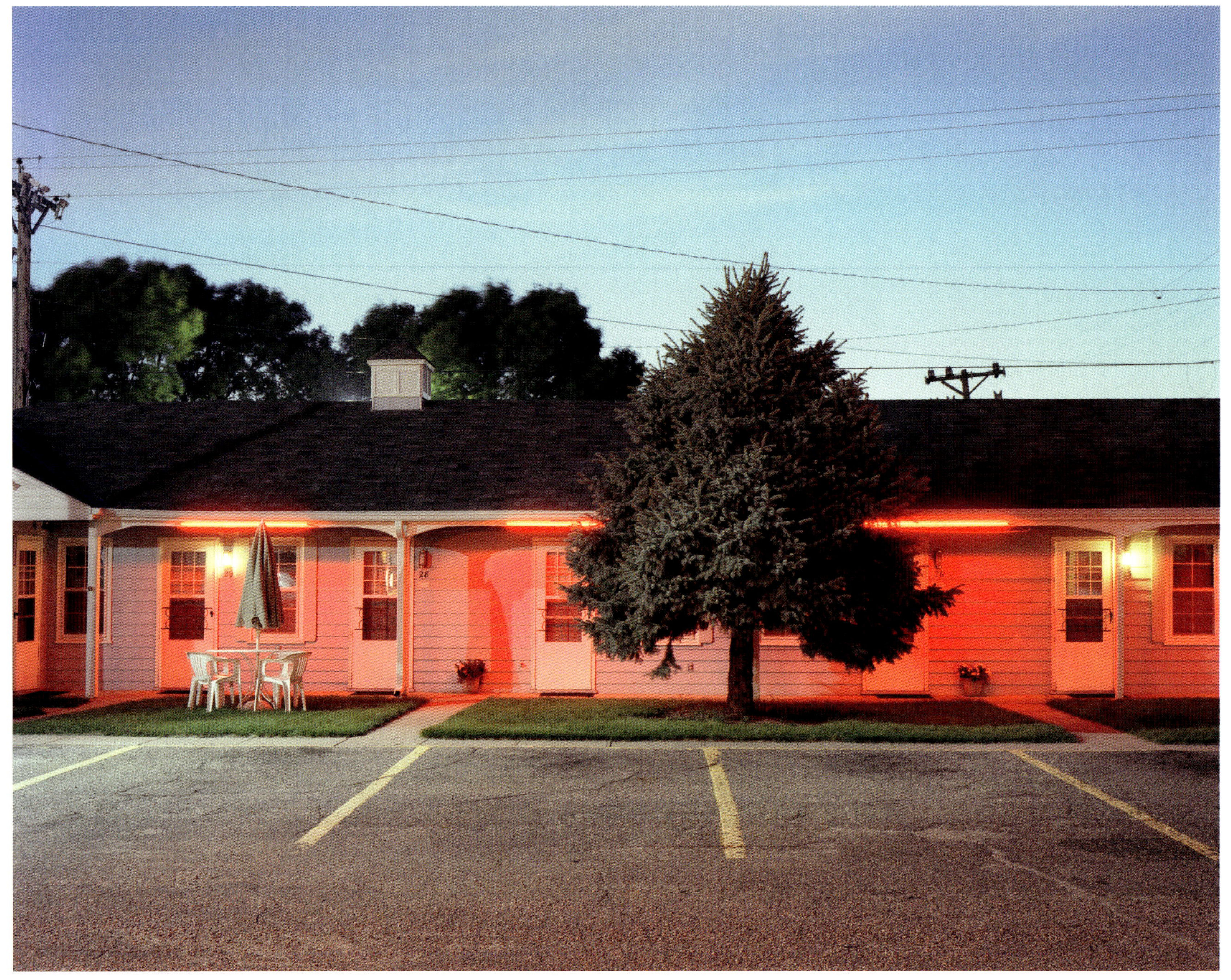

The Colonial Inn, Yankton, South Dakota

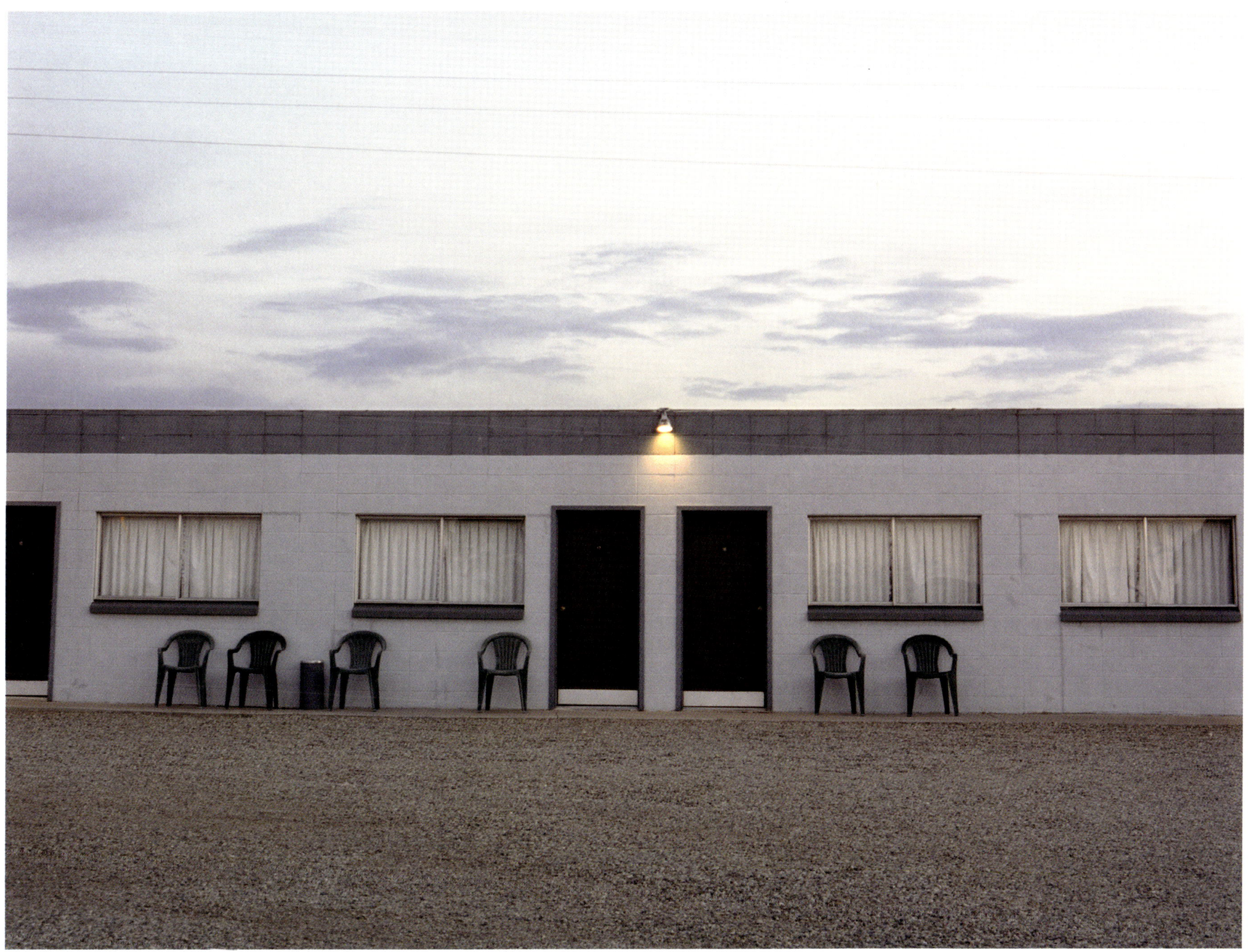

The Best Value Inn, Hanksville, Utah
The Butch Cassidy gang allegedly hid out in the canyons east of town

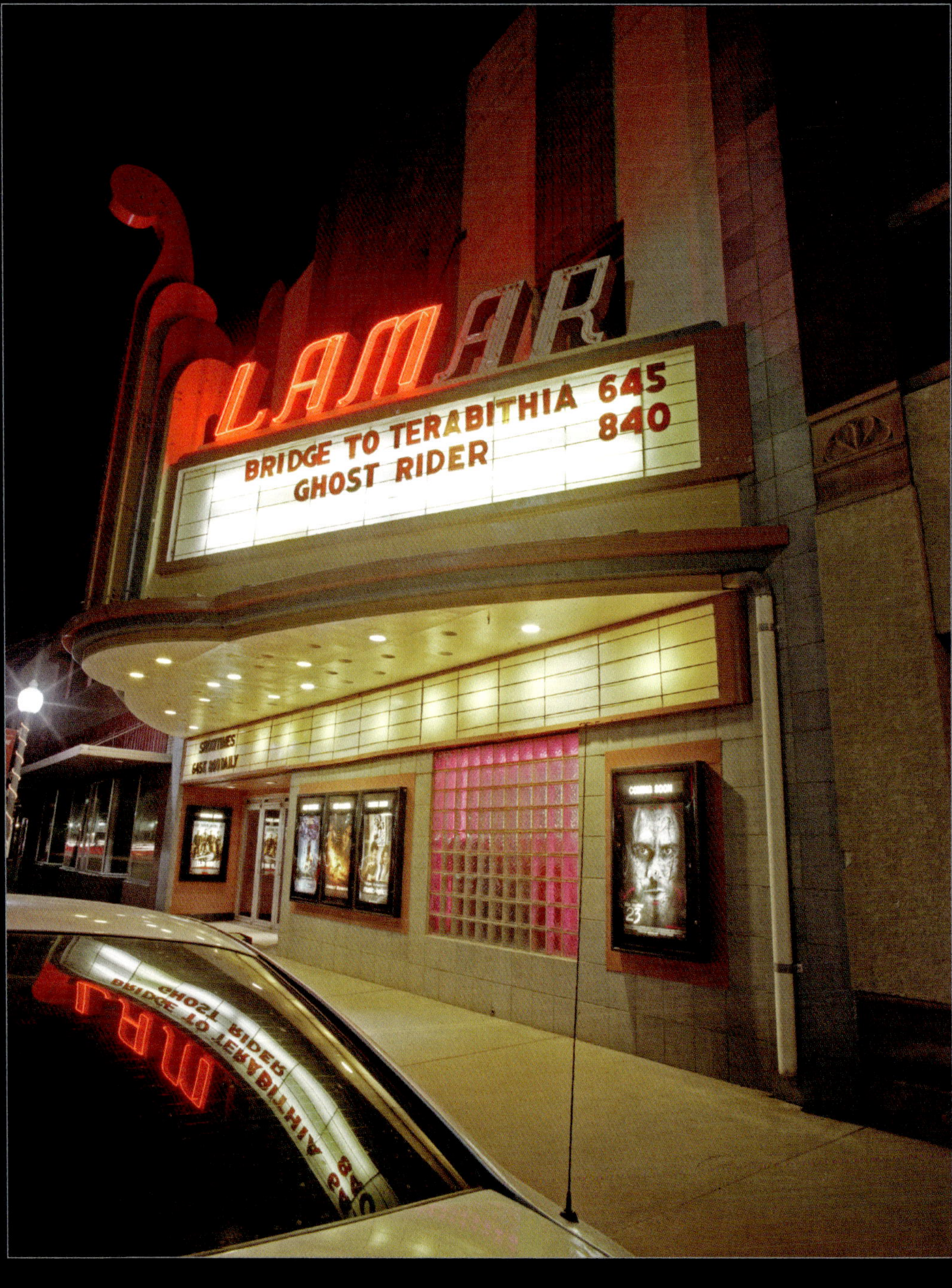

The Lamar Theatre, Lamar, Colorado

A wonderful Art Deco exterior which has been lovingly restored and preserved

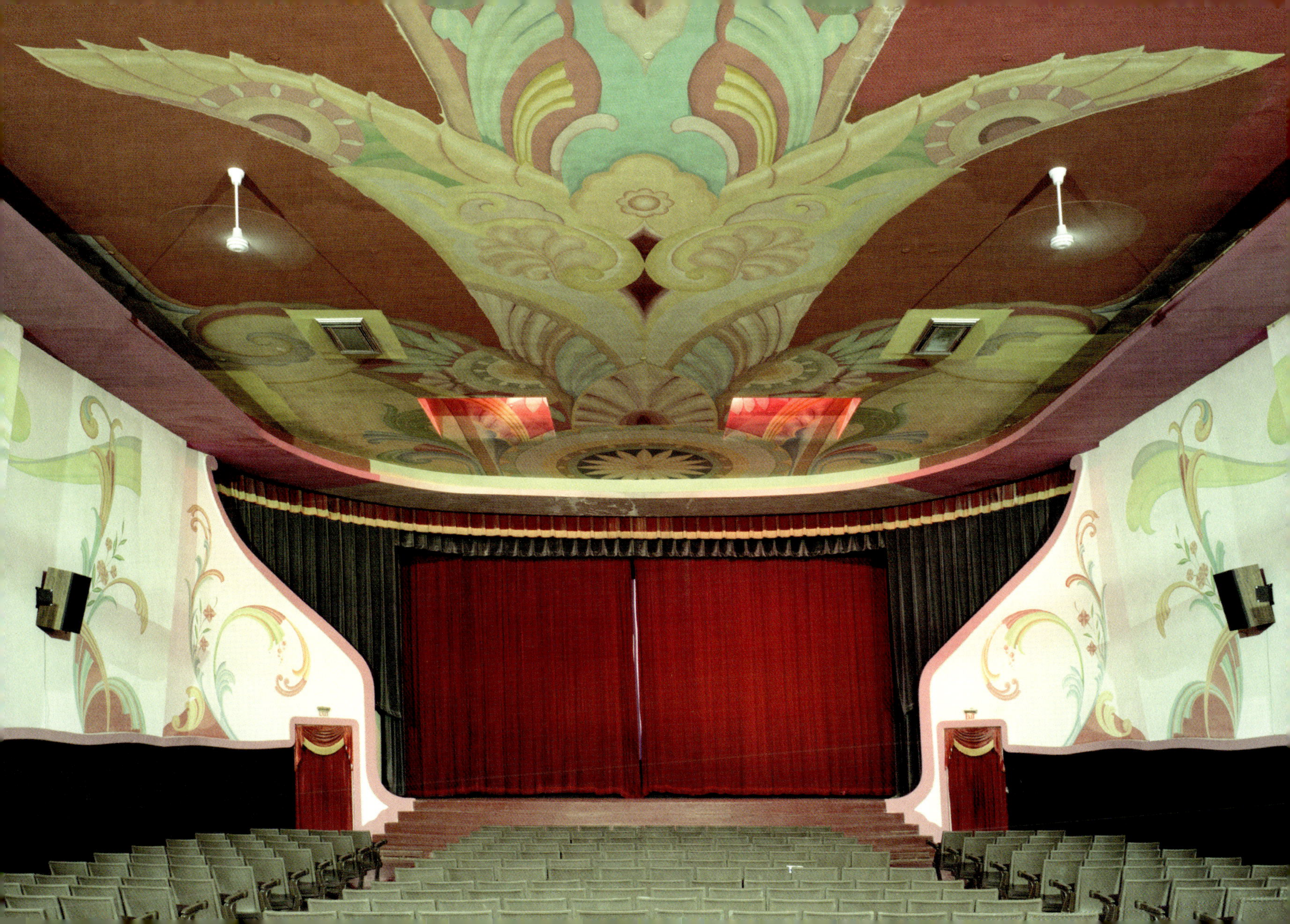

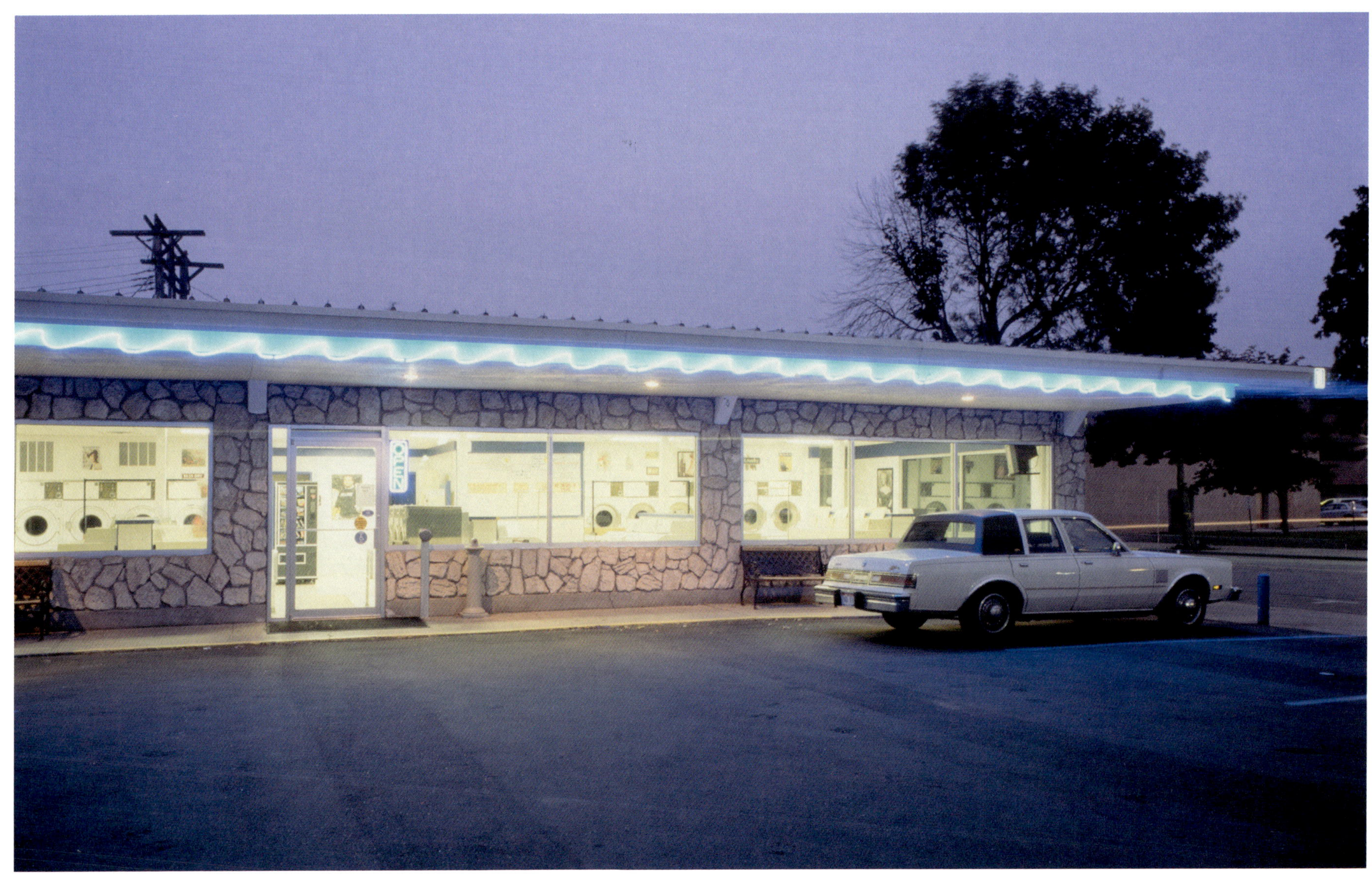

Laundromat in early morning light, Yankton, South Dakota

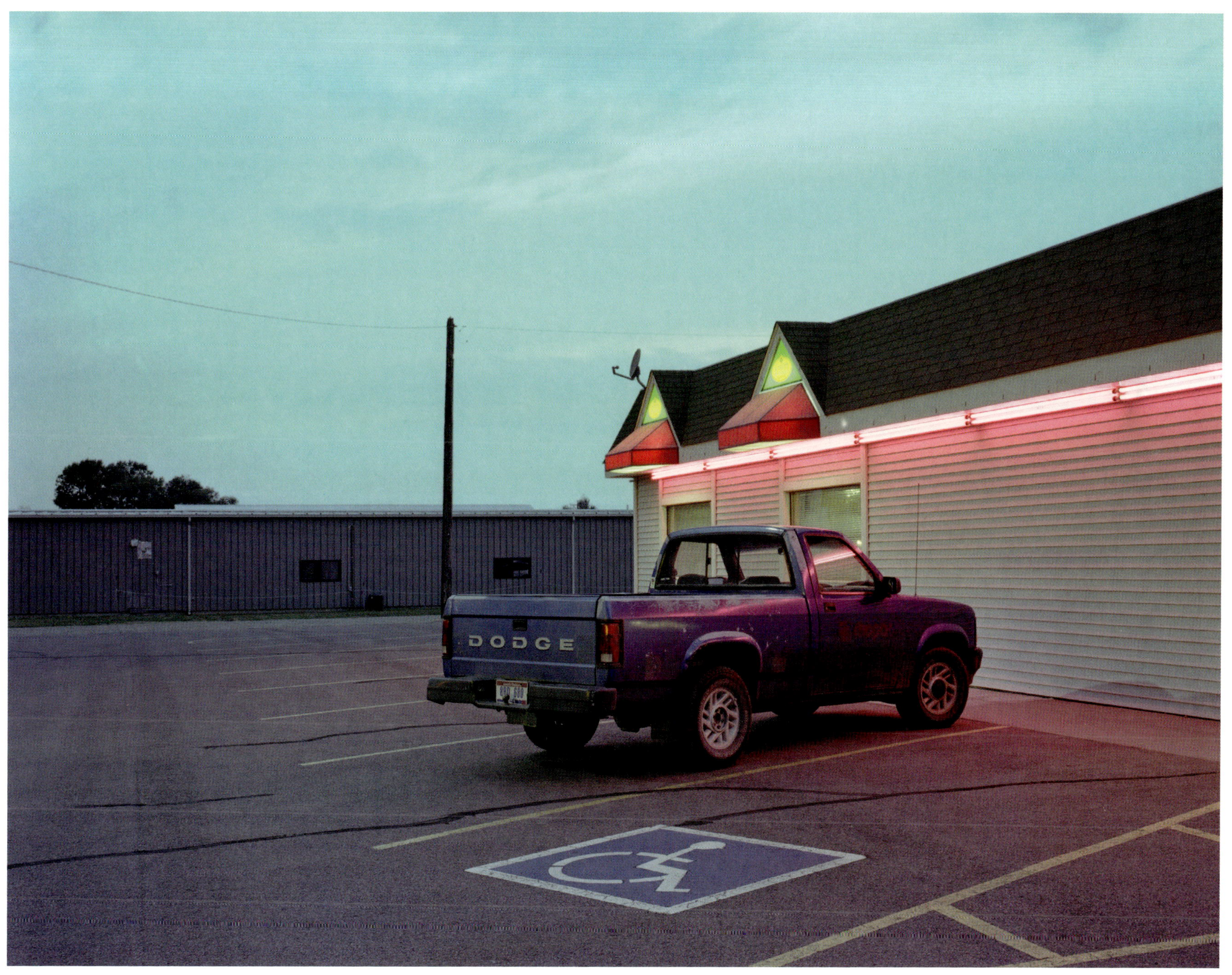

Dawn at The Country Restaurant, Mitchell, South Dakota

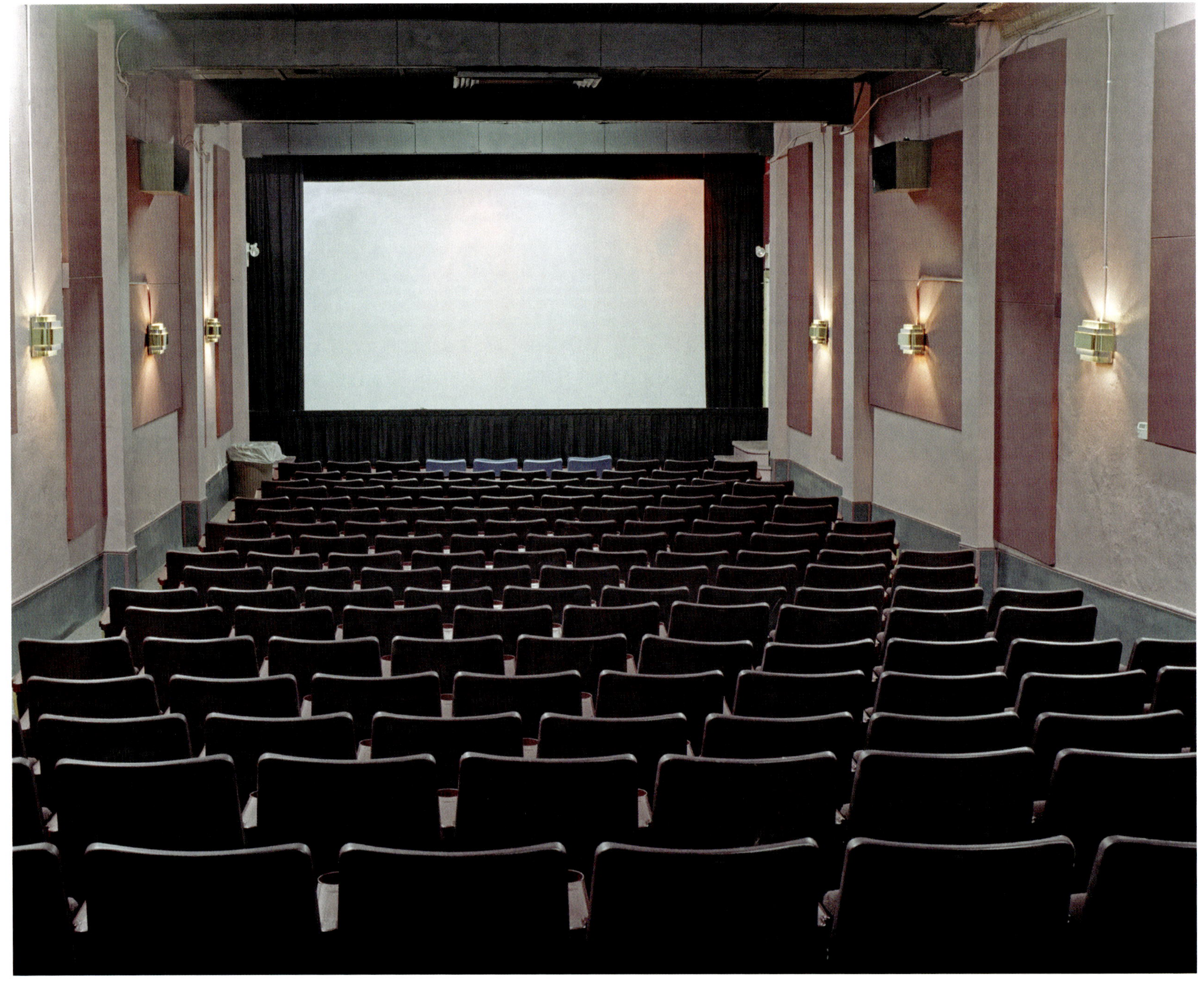

Interior of the Holly Theater, Holly, Colorado

Amazingly, with a population of only 500, Holly is still able to retain its movie theater thanks to the commitment of volunteers

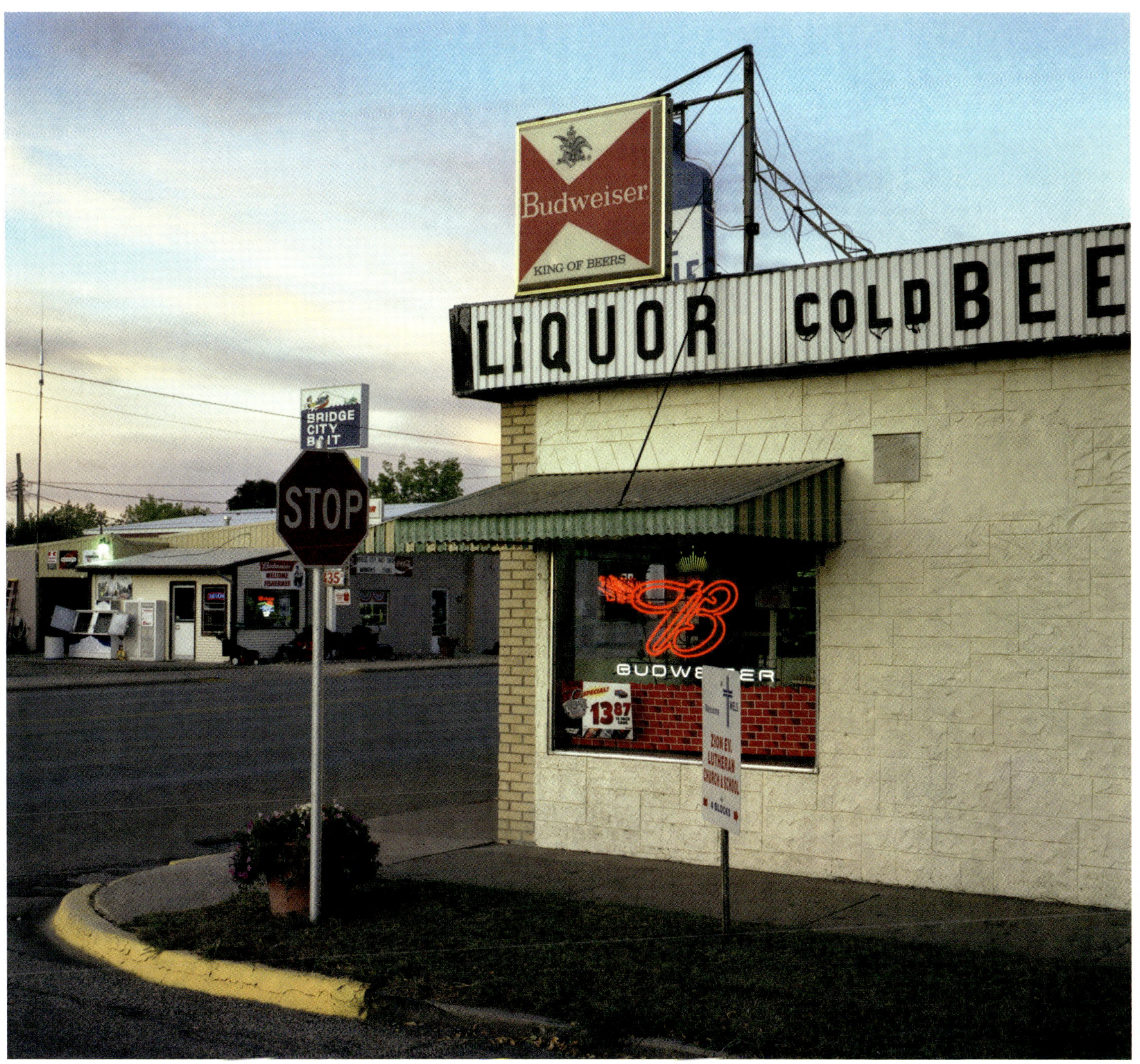

Liquor store at dawn, Mobridge, South Dakota

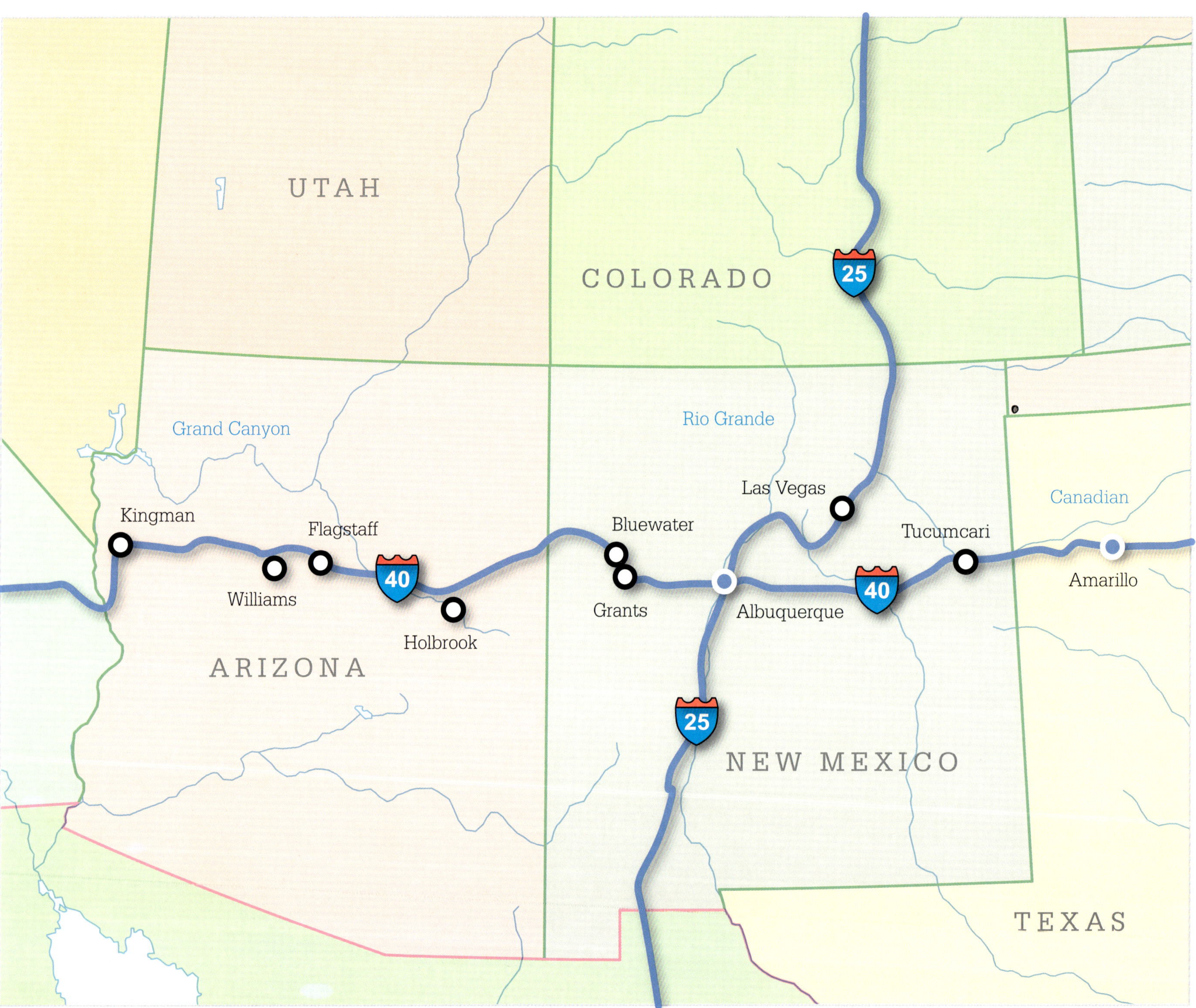
UTAH
COLORADO
25
Grand Canyon
Rio Grande
Las Vegas
Canadian
Kingman
Bluewater
Tucumcari
Flagstaff
40
Williams
Grants
Albuquerque
40
Amarillo
Holbrook
ARIZONA
25
NEW MEXICO
TEXAS

Route 66, through New Mexico and Arizona

Route 66 is unquestionably one of the most famous highways in the world, no doubt through being immortalized by John Steinbeck in *The Grapes of Wrath* (1939). It stretches from the Pacific Coast to Chicago in the Midwest; as it crosses Arizona and New Mexico it passes through some of the country's most iconic towns, rich with cafes and motels and a style of Americana one immediately associates with the 1940s and 1950s. This is a land of billboards, statues, and kitsch.

Sadly, Route 66 is in decline – and has been for decades. The rot started in the late 1950s when a network of high-speed interstates was created to bypass the towns. When the last stretch of Interstate 40 was completed in 1984, Route 66 was officially decommissioned, being re-designated Historic Route 66. While the developments may have been welcomed by some motorists, they created immediate economic challenges for those businesses left along the old highway. Today, Route 66 is of semi-mythological status, a legend some of whose reality still abides. Even today you can drive on parts of the original route, particularly through the communities that once served it, but increasingly the stretches of highway between the towns are disappearing. You might travel a dozen miles or so, then quite without warning find yourself confronted by a red-and-white-striped No Entry barrier or forced onto the interstate, from where you can see the original road emerge then disappear without trace, only to reappear once more alongside an abandoned gas-station.

What marked out Route 66 was a distinct entrepreneurial spirit, as manifested by the numerous advertising billboards and statues constructed along the highway to tempt the passing motorist toward a particular motel or restaurant. Fortunately this spirit is still very much alive. The towns that were bypassed by the new interstate are appealing to motorists' sense of nostalgia and history in an attempt to lure them back. These communities offer a wonderful insight into a vanishing American past, and they are not reluctant to point this out, with most motels and cafes playing up their Route 66 connections. Even so, you're left wondering just how long this can last.

Arizona offers several of the best surviving stretches of the old road. Starting from the Colorado River, the route passes through dozens of wonderful old highway towns. On leaving the California border, it cuts across a particularly unforgiving stretch of desert until finally reaching the former gold-mining town of Oatman. Eastward from here it follows a steep incline over the Black Mountains and into the lively railway town of Kingman. Kingman is well served commercially, being at the confluence of Historic Route 66 and Interstate 40; it retains many of the old motels and diners, which draw trade from both sources. Continuing out of Kingman to pass through the Hualapai Valley, Route 66 hits a particularly evocative stretch where the road runs parallel to the Santa Fe Railroad tracks, passing a collection of all but forgotten towns, the most interesting being Hackberry, home to the Old Route 66 Visitor Center. Beyond Hackberry are various other communities that make much of their Route 66 associations in their attempts to entice you, including Seligman and Williams, a town rich in neon lighting which fabulously evokes the 1940s and 1950s.

As good fortune would have it, this part of the highway passes close to the Grand Canyon, so attracting tourists is less of a problem than it might have been. Nearby Flagstaff is a vibrant town, home to the Northern Arizona University. Surrounded by mountains – atop one of which is one of the most famous astronomical observatories in the world – Flagstaff is a mecca for mountain bikers and skiers and is a good striking-out point for some of the nation's notable attractions, including the Barringer Meteor Crater, so has been little affected by the downgrading of Route 66. Further east the route rises to the Kaibab Plateau, but then steadily descends toward a dryer, more desert-like landscape.

While little of the original road exists in New Mexico, many of the old towns remain. The landscape in the western part of the state is particularly appealing, with its looming sandstone mesas. Gallup, the first community of any note, is a classic railroad town, its main commercial areas running parallel to the railway. Sandwiched between the Navajo and Zumi reservations, it also reflects a broad cultural diversity. The town of Grants was originally developed for uranium mining, but has retained much of its original Route 66 charm. In common with other smaller communities like Santa Rosa and Tucumcari it continues to survive mainly because the new interstate passes close by. Tucumcari deserves special mention as it still has many of the features that made it such an attractive stopping-off point in its Route 66 heyday; the Blue Swallow Motel there must be one of the best preserved establishments along the entire route, and has featured in TV programs as well as in *Smithsonian Magazine*.

As you continue eastward toward the Texas border the landscape becomes flatter but still seems to be filled with a wonderful sense of freedom.

Affectionately referred to as the Mother Road, this fabled highway will always be close to the hearts of those who enjoy traveling across America. The interstate, coupled with cheap inter-city flights, may have rendered Route 66 obsolete, but it is hard to imagine that it will not continue to inspire new generations of novelists and songwriters. And a single unforgettable line by Bobby Troup, the former pianist with the Tommy Dorsey Band, has perhaps single-handedly ensured this wonderful highway's immortality in American popular culture: "Get your kicks on Route 66."

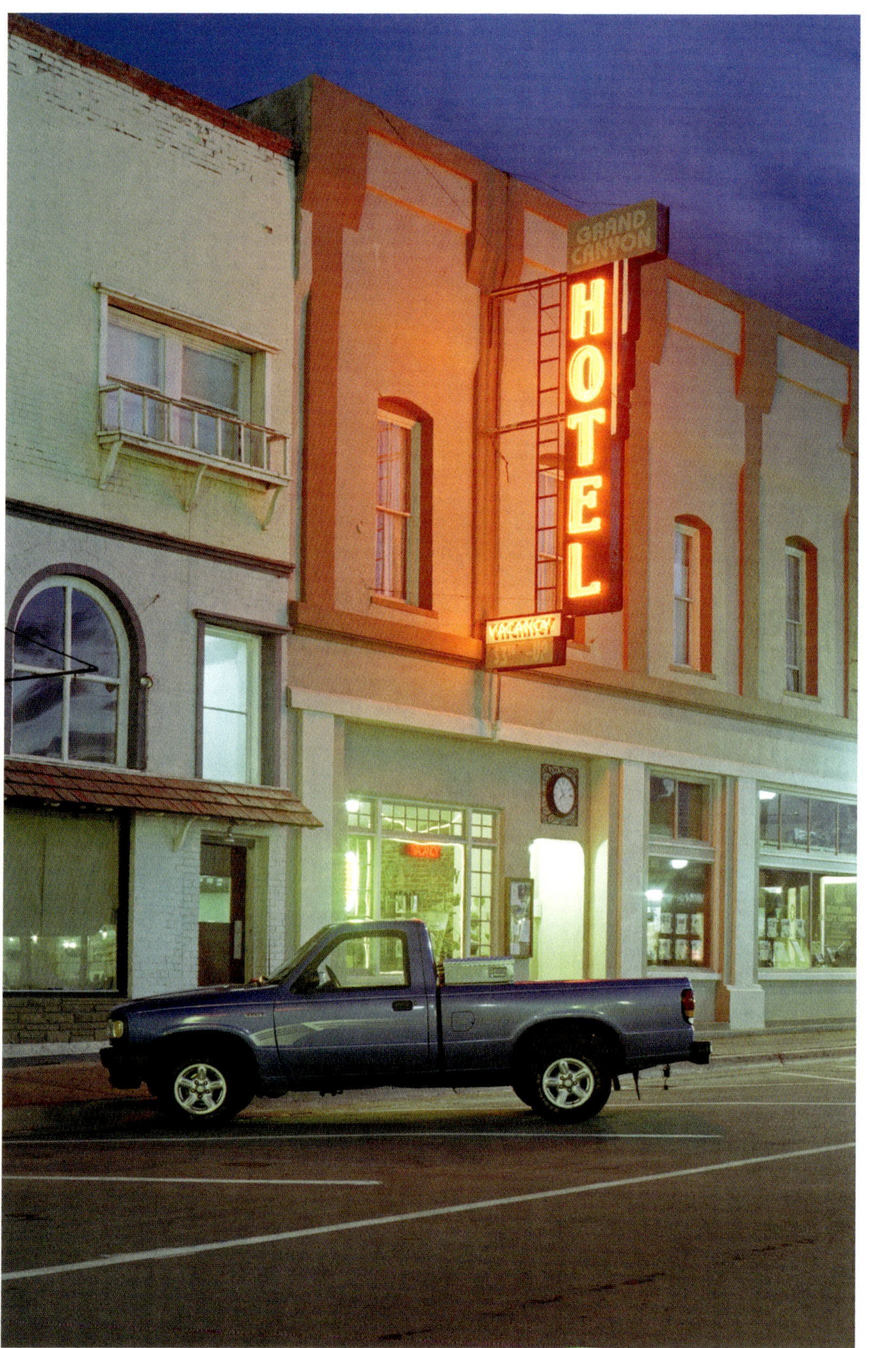

Hotel, Williams, Arizona

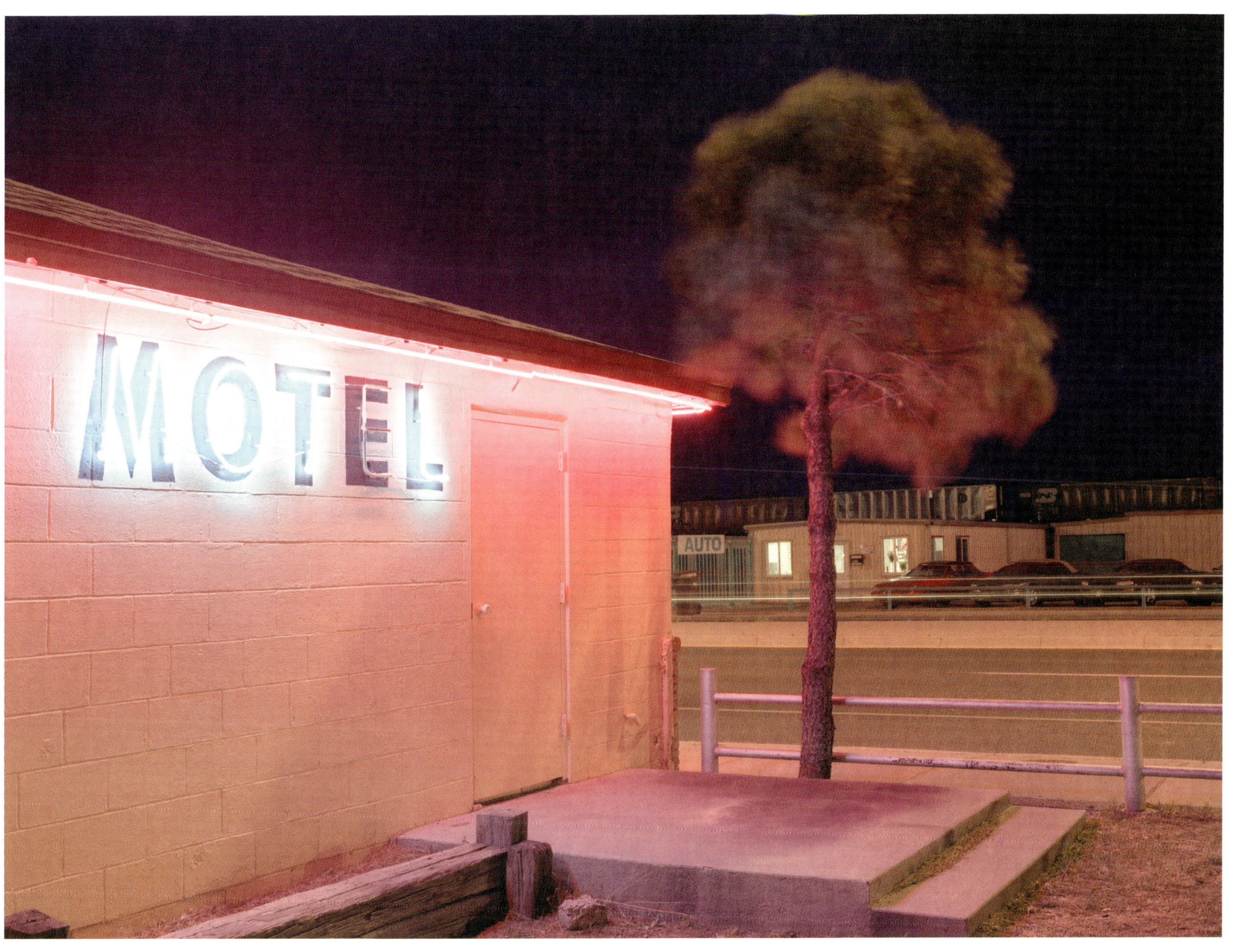

Fluorescent motel, Grants, New Mexico

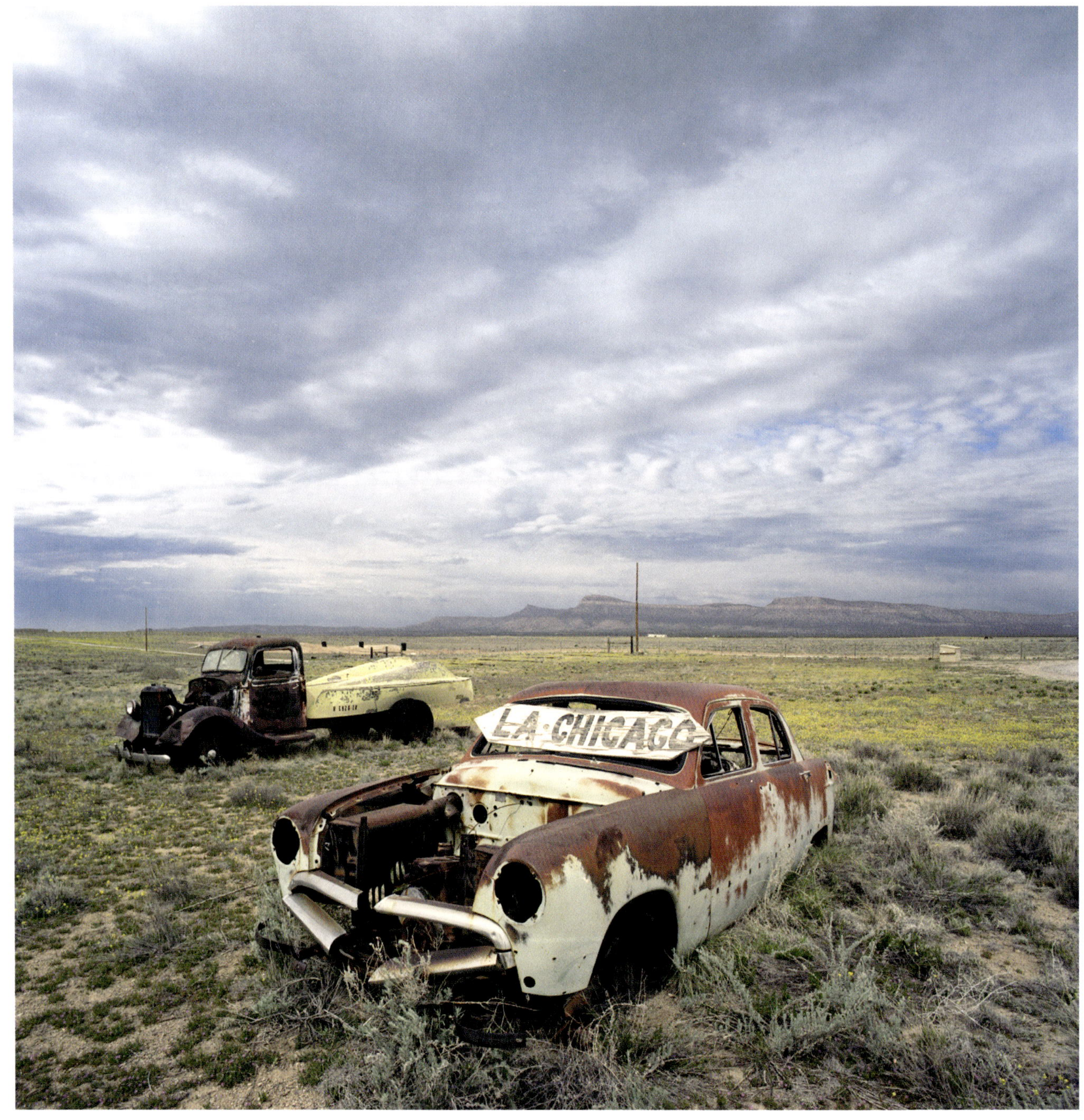

Abandoned vehicles on Route 66, west of Kingman, Arizona

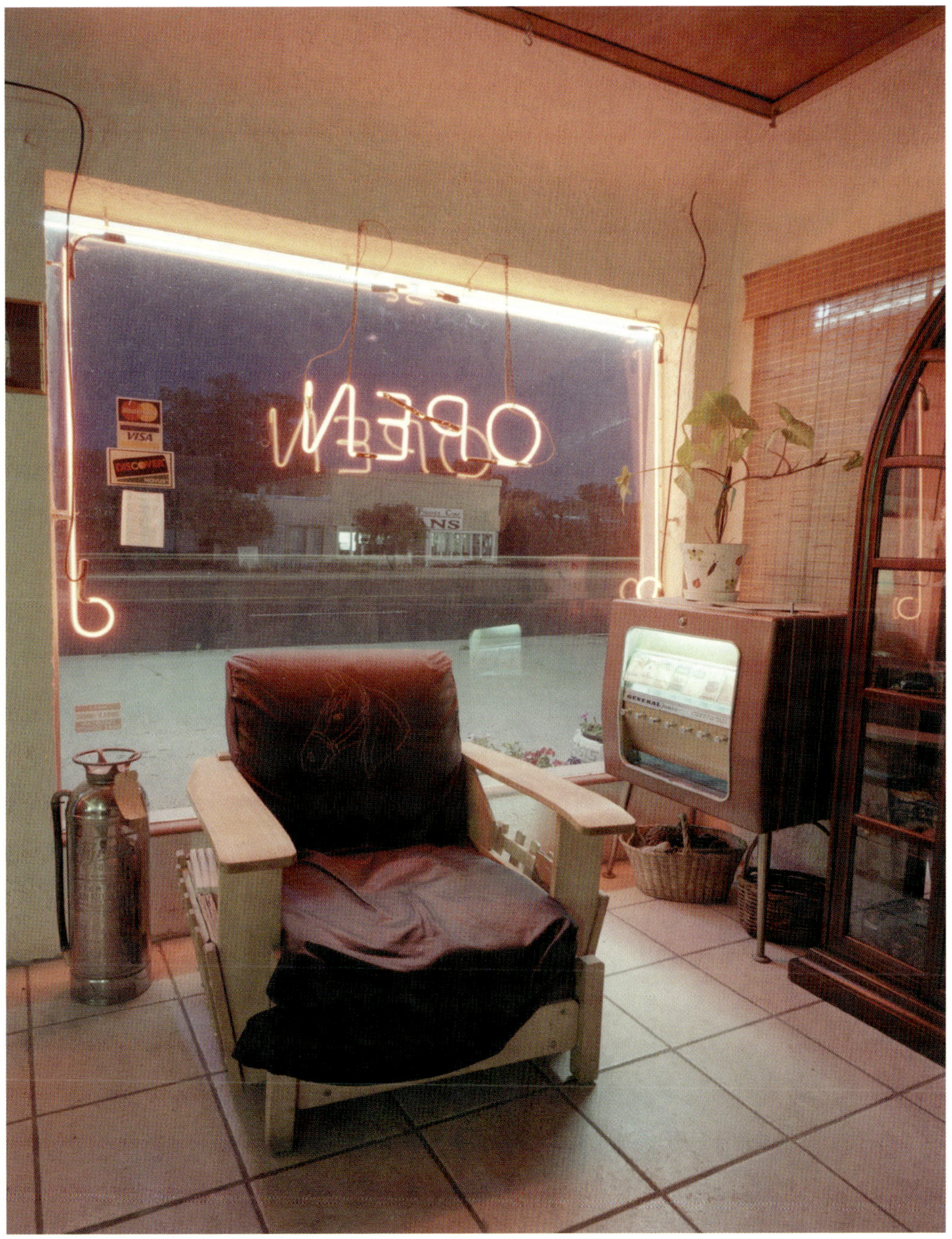

Foyer, the Blue Swallow Motel, Tucumcari, New Mexico
The Cozy Cone Motel in the cult movie "Cars" is allegedly partially based on this famous Route 66 institution

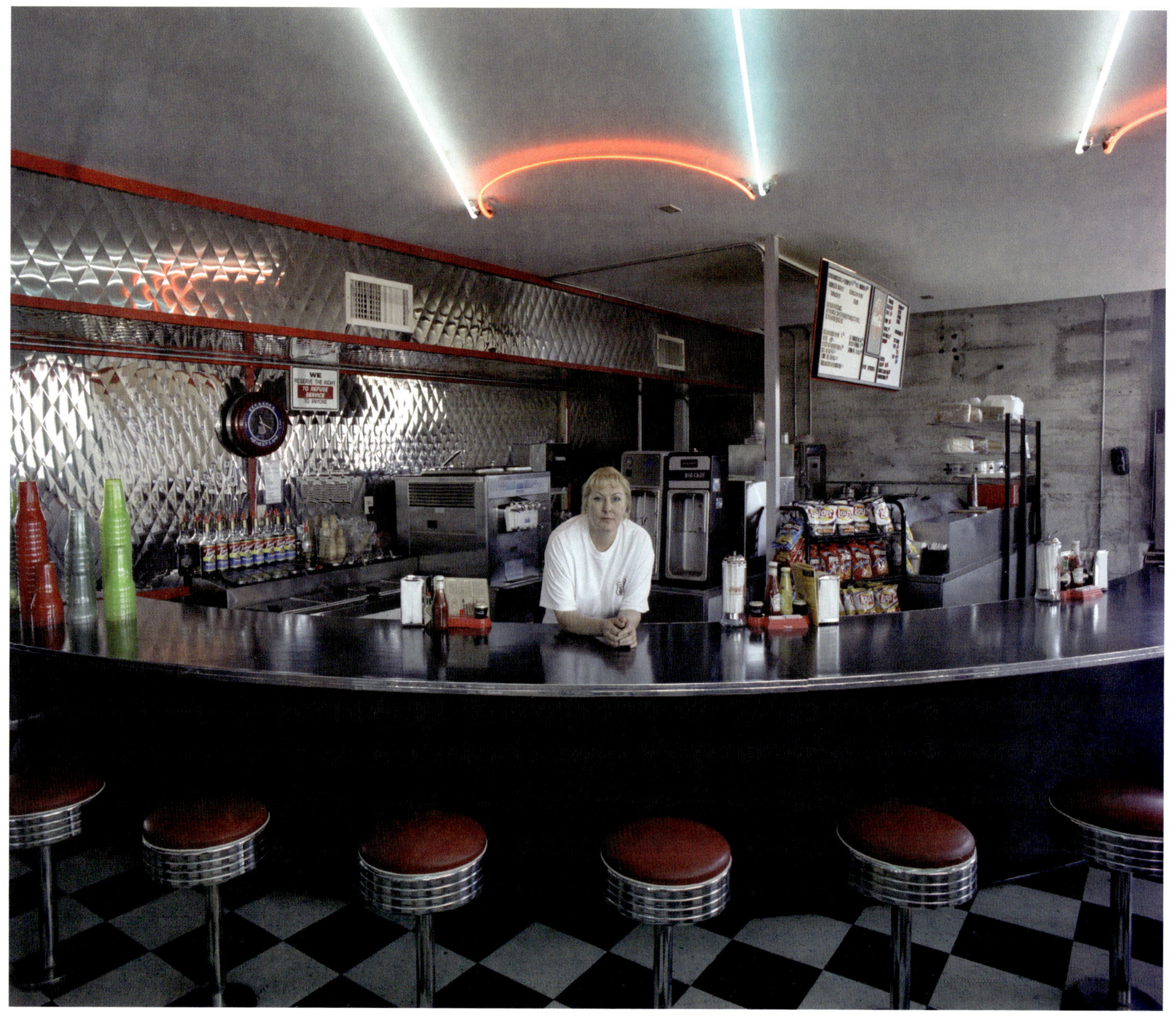

Diner, Kingman, Arizona

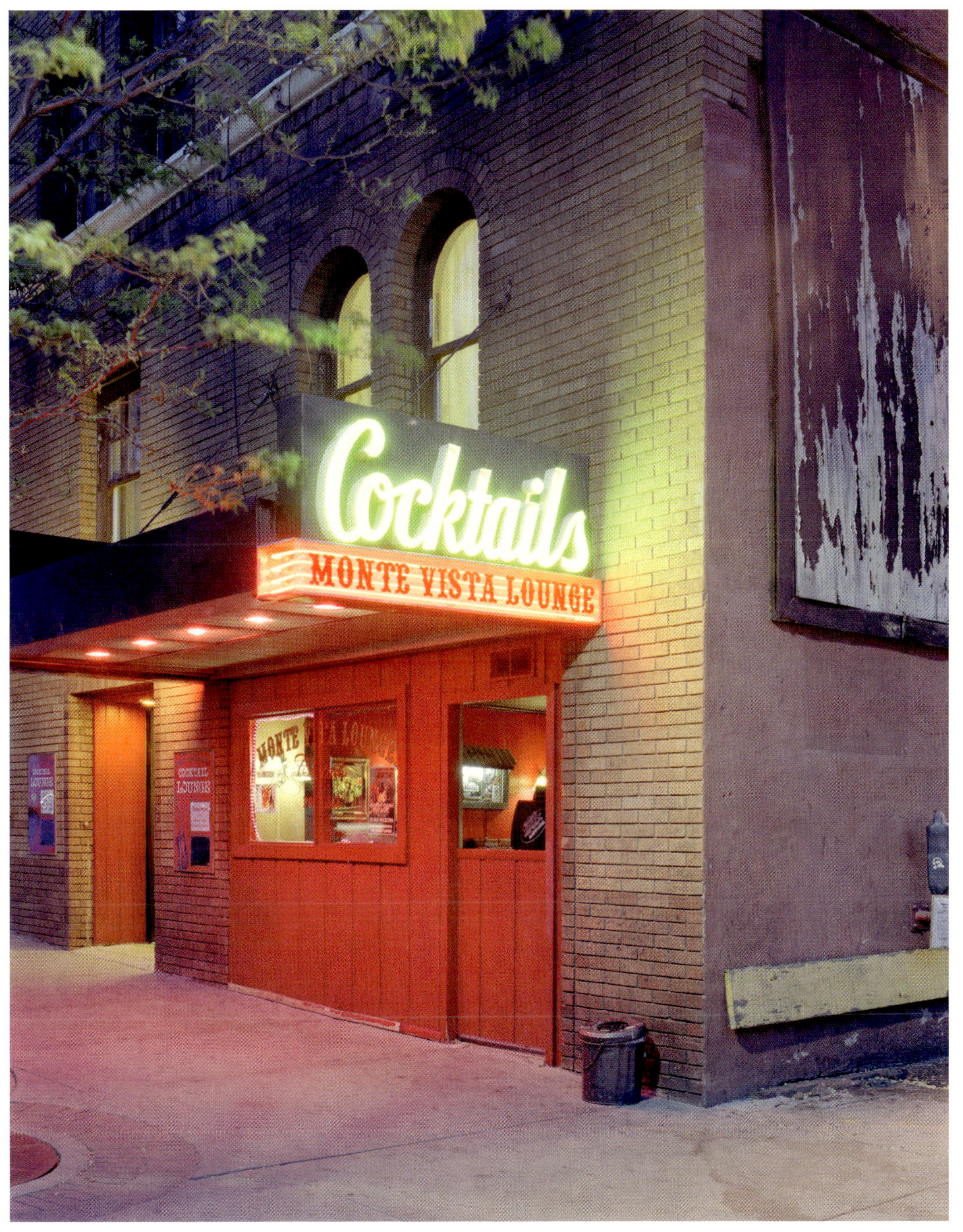

Cocktail bar, Monte Vista Hotel, Flagstaff, Arizona
A famous haunt for stars, scenes from the classic movie "Casablanca" were filmed here

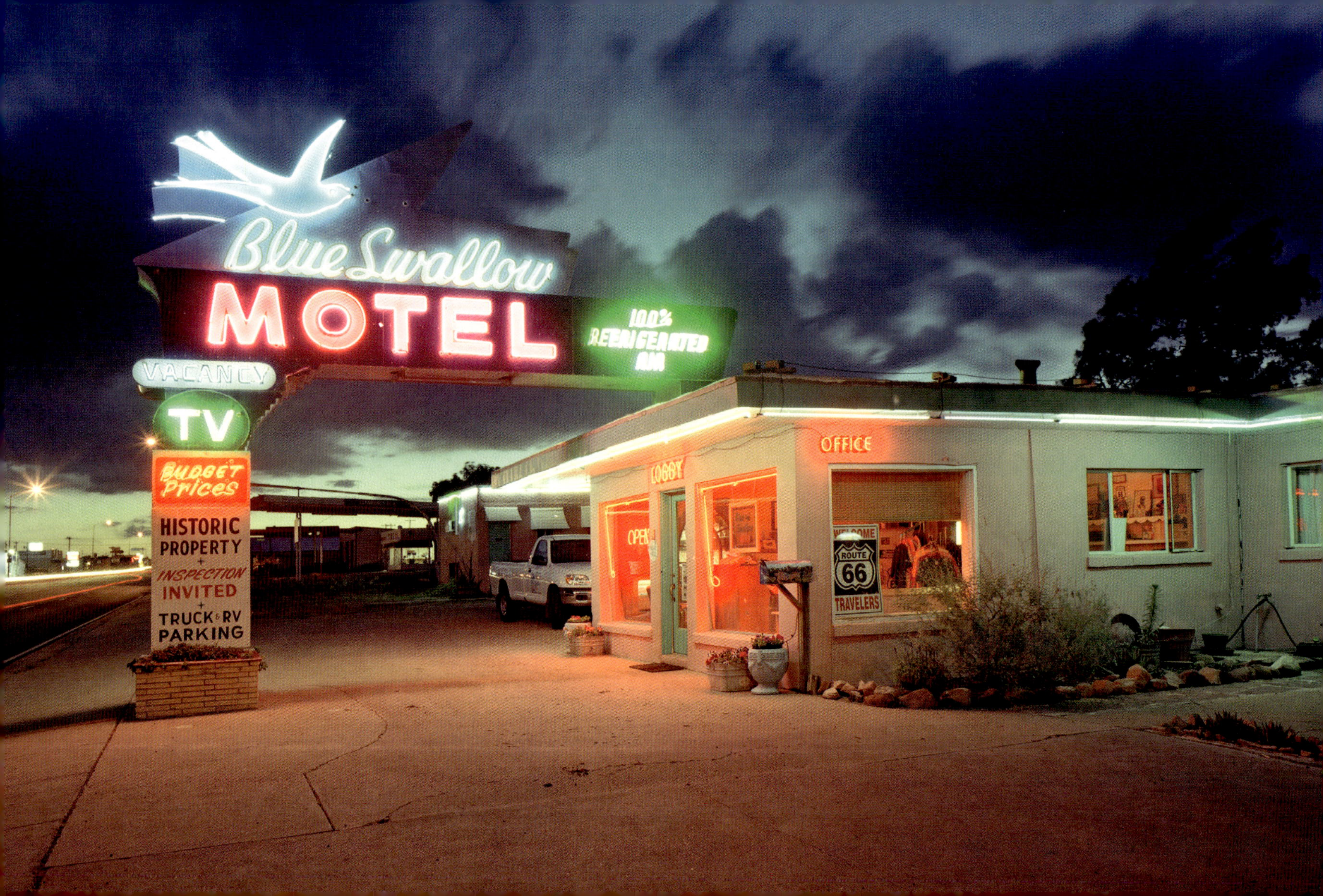

Blue Swallow
MOTEL
100% REFRIGERATED AIR
VACANCY
TV
Budget Prices
HISTORIC PROPERTY
+
INSPECTION INVITED
+
TRUCK & RV PARKING
OFFICE
LOBBY
OPEN
WELCOME
ROUTE
66
TRAVELERS

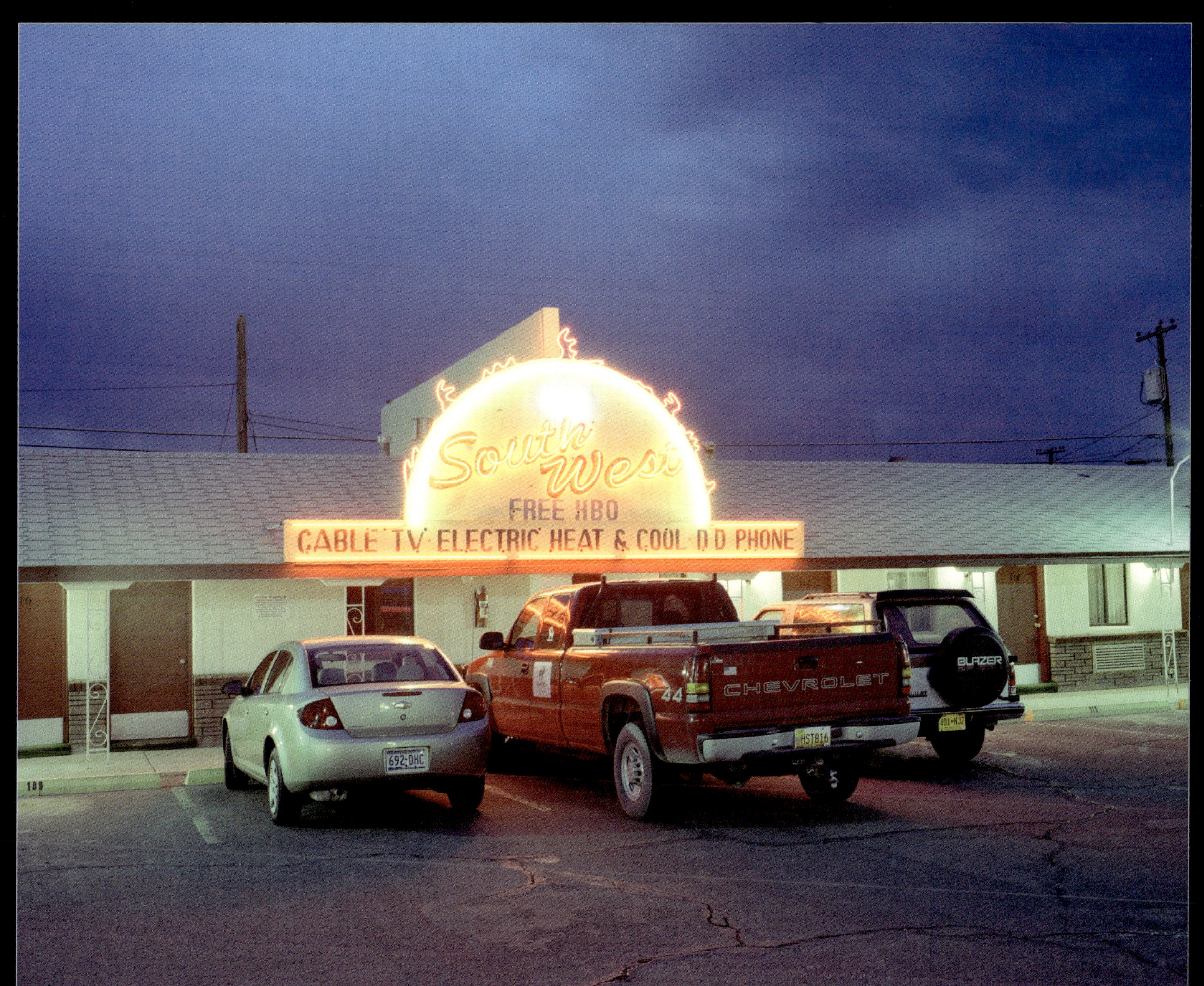

South West
FREE HBO
CABLE TV ELECTRIC HEAT & COOL D D PHONE
CHEVROLET
BLAZER
692 DHC
HST816

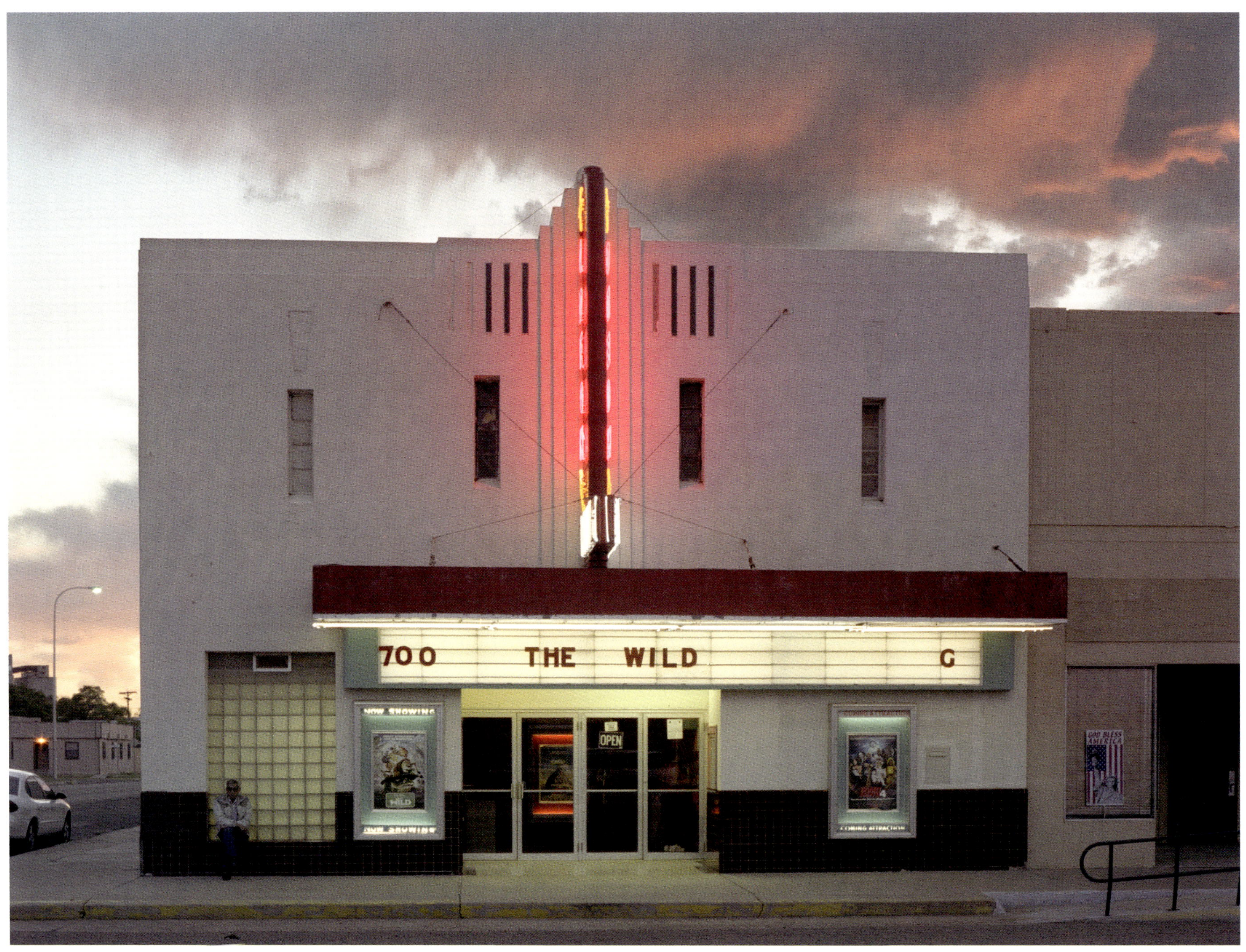

The Odeon, Tucumcari, New Mexico

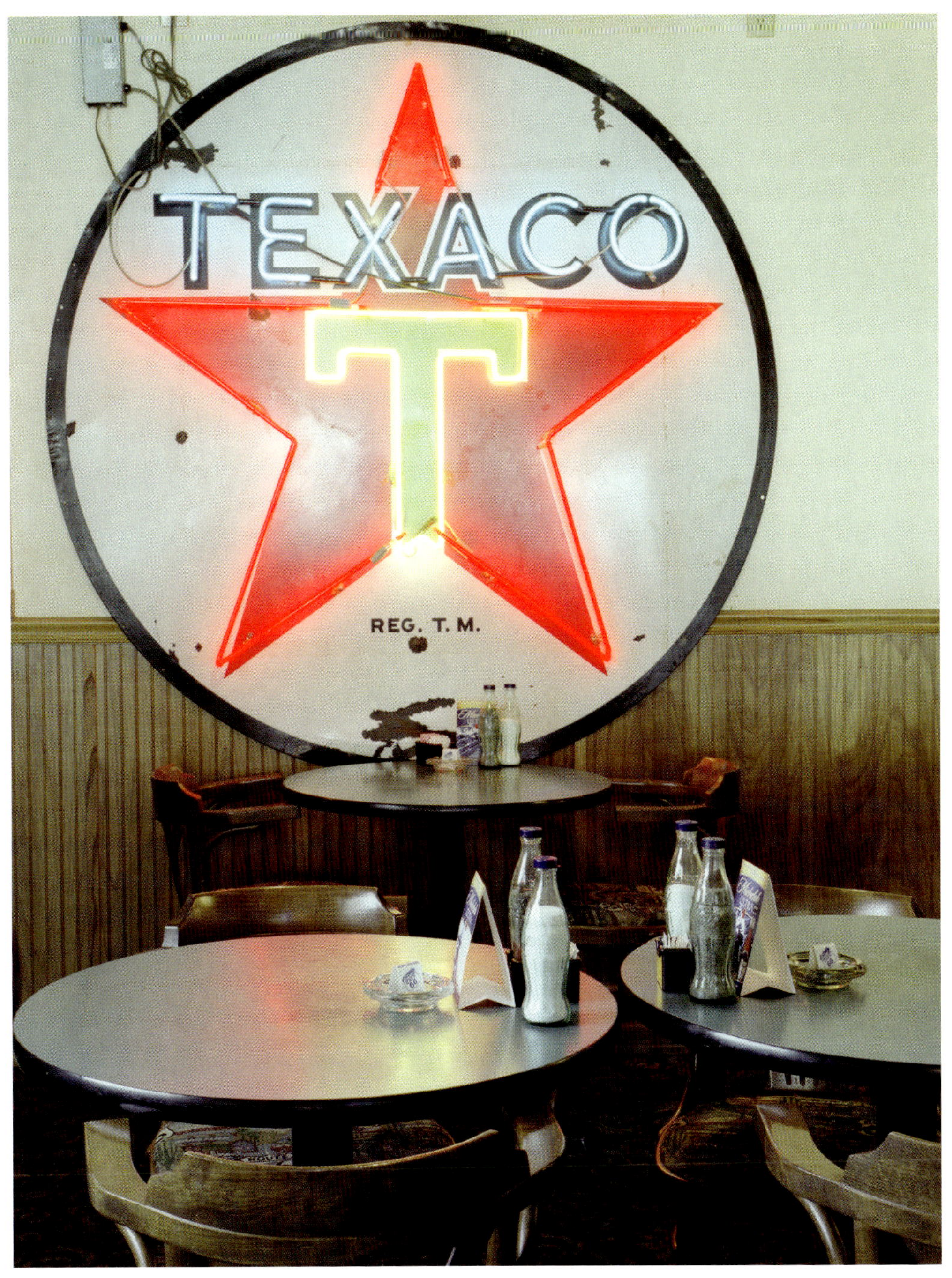

Cafe interior, Williams, Arizona

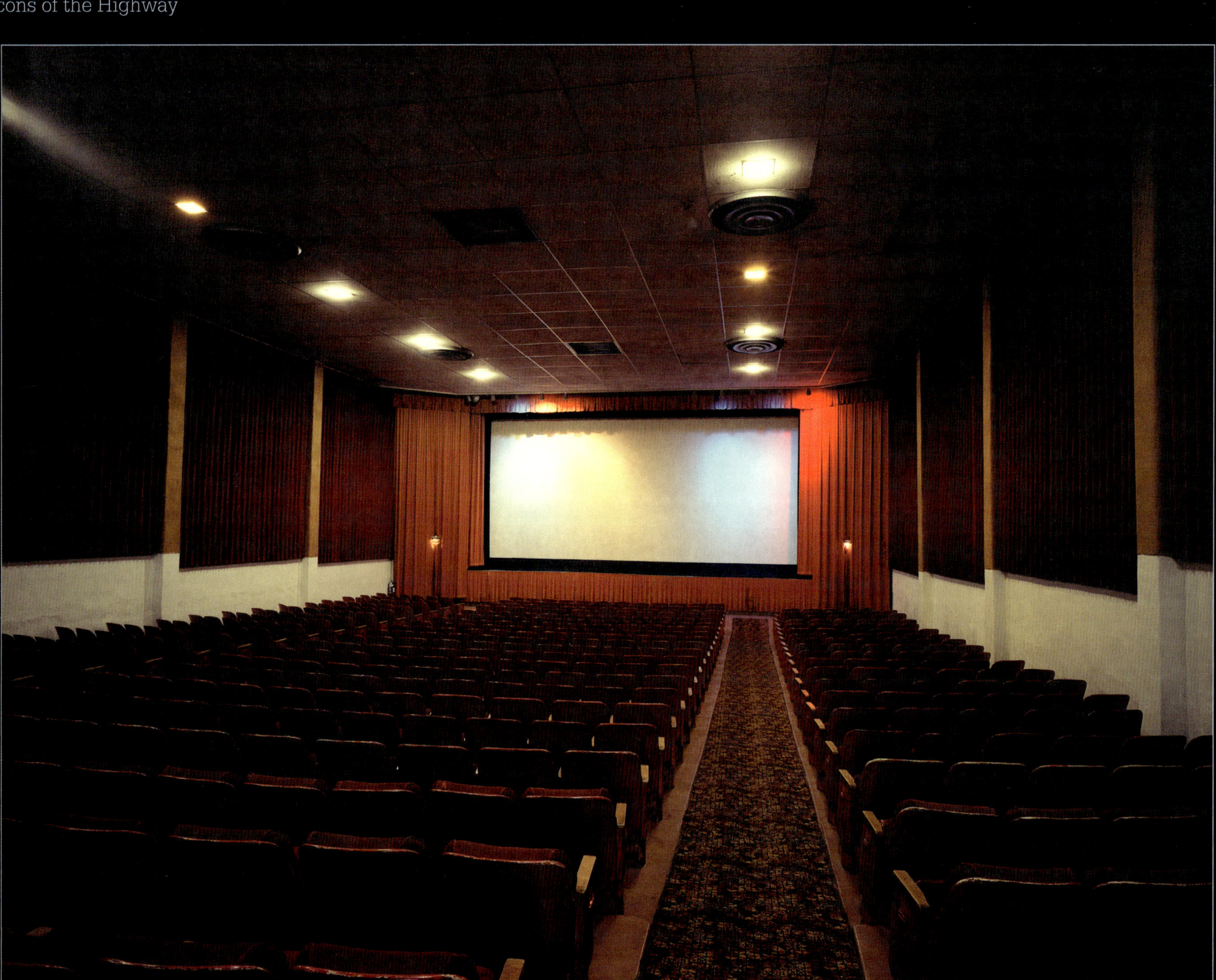

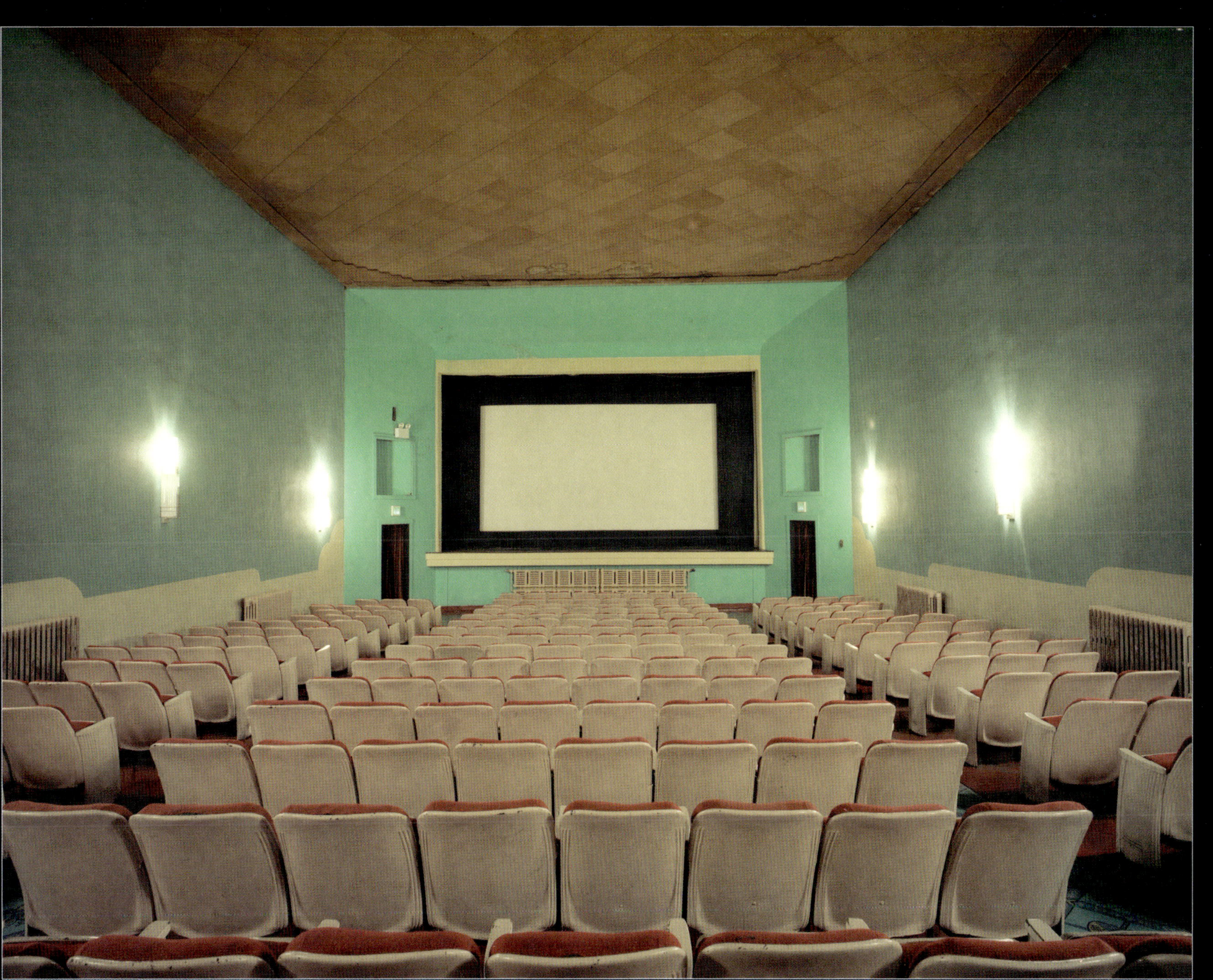

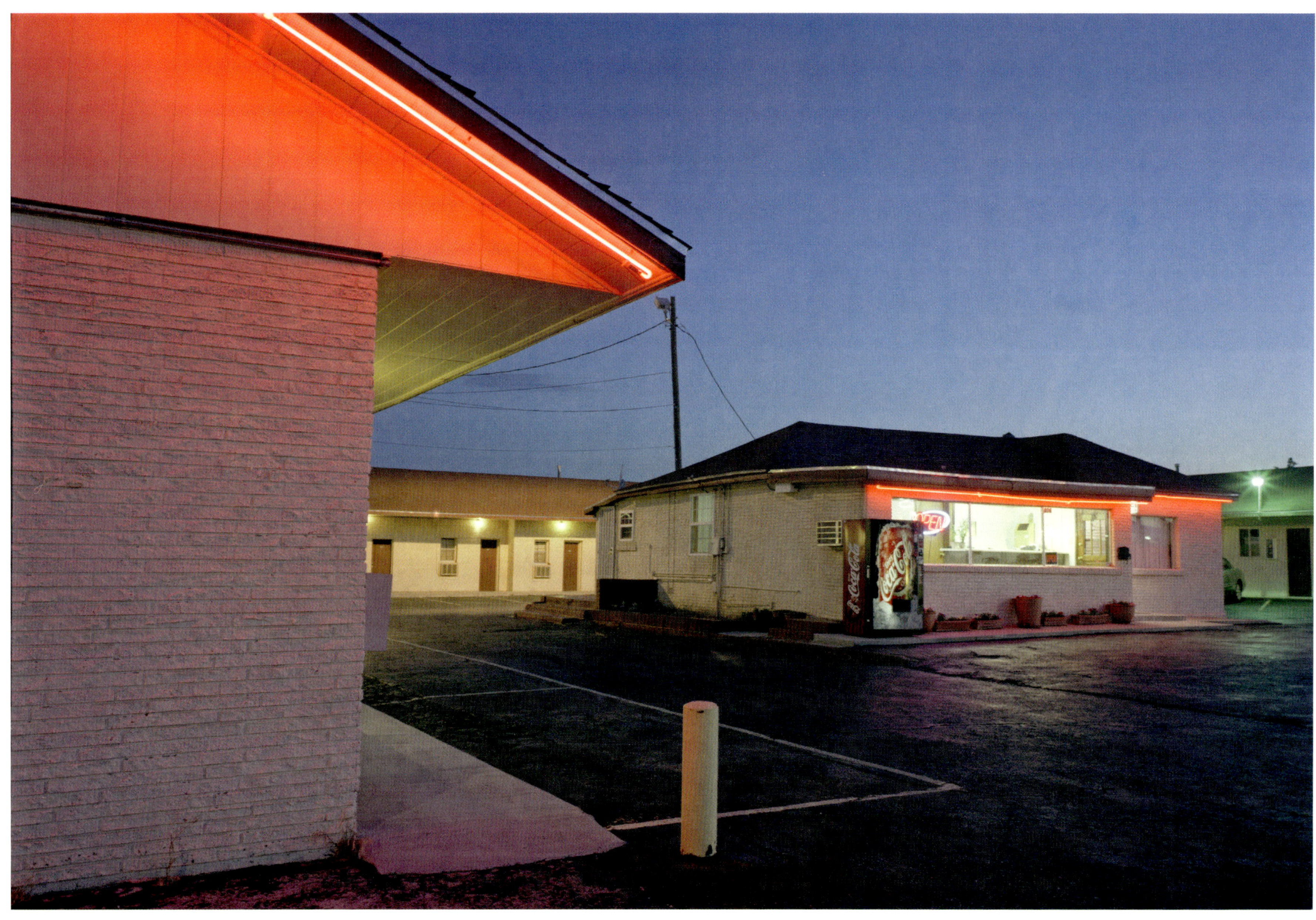

The Americana Motel, Tucumcari, New Mexico

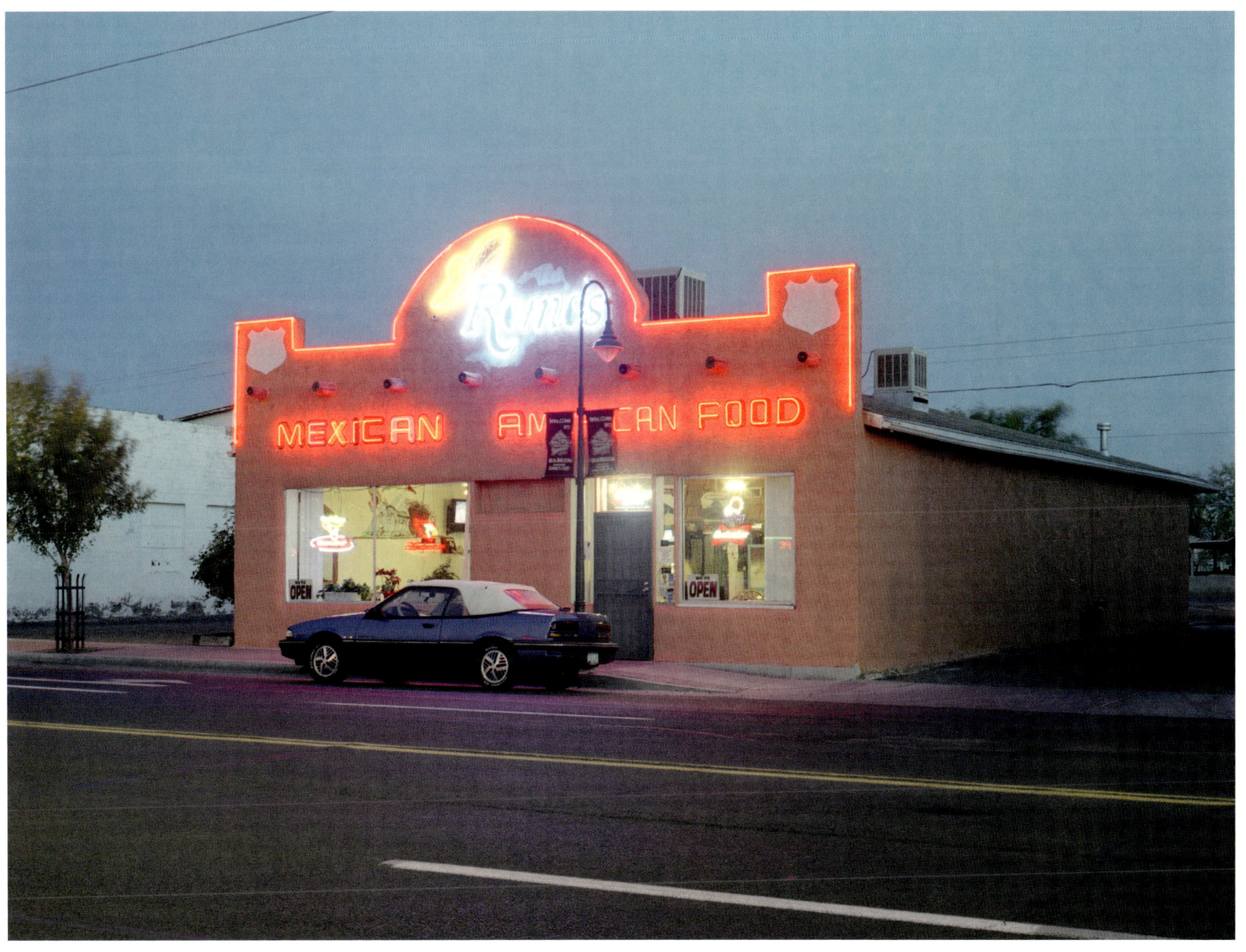

Romo's Restaurant, Holbrook, Arizona

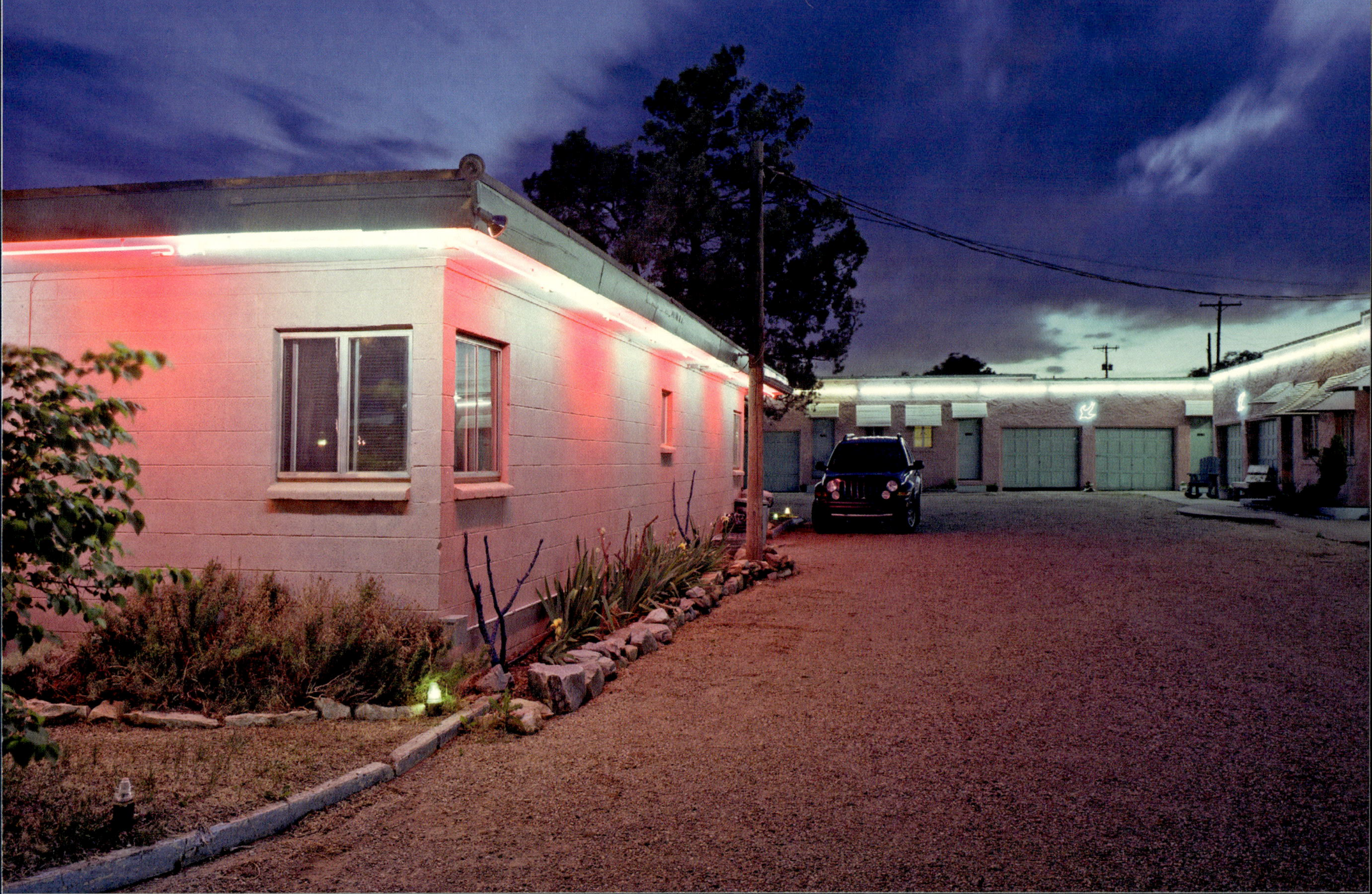

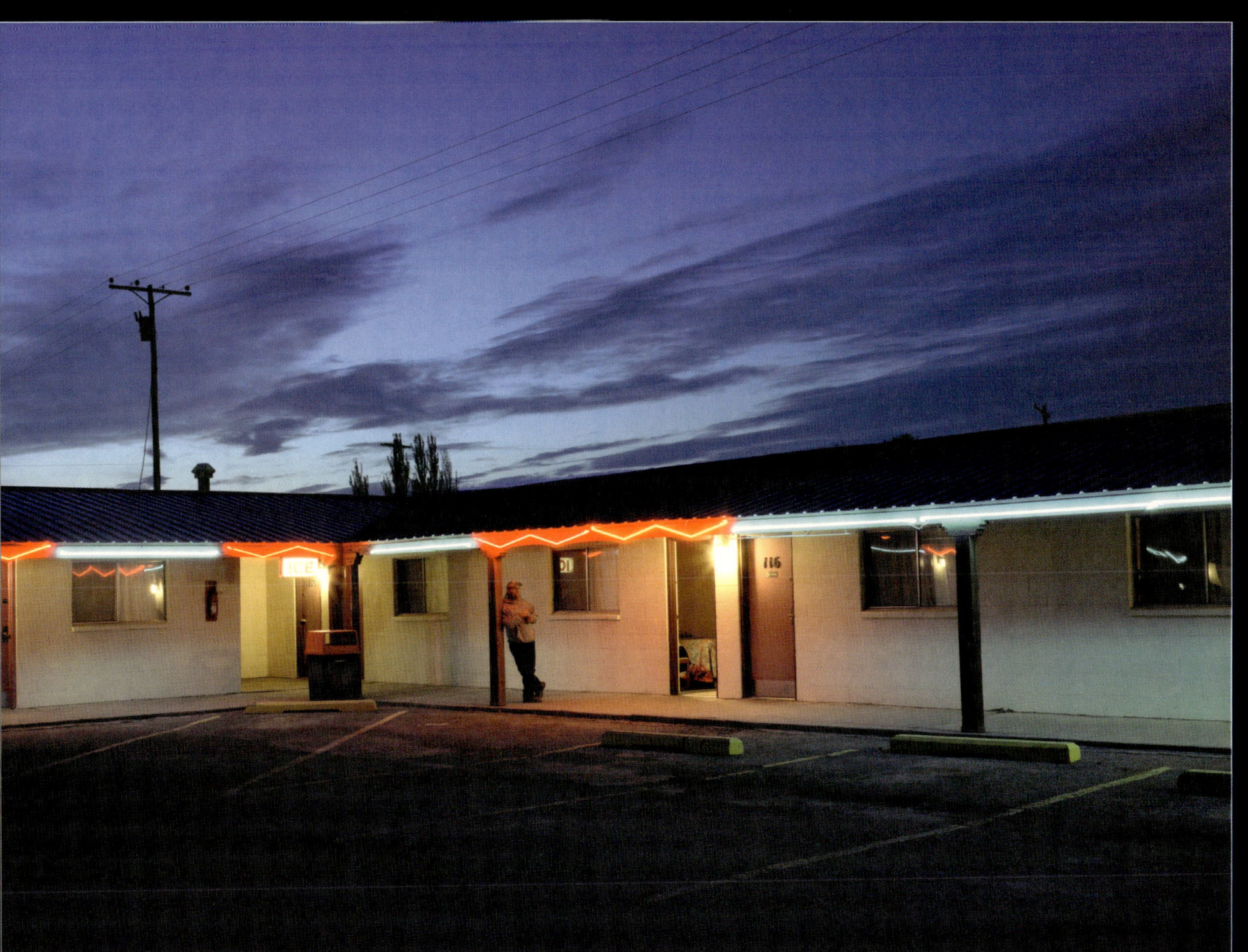

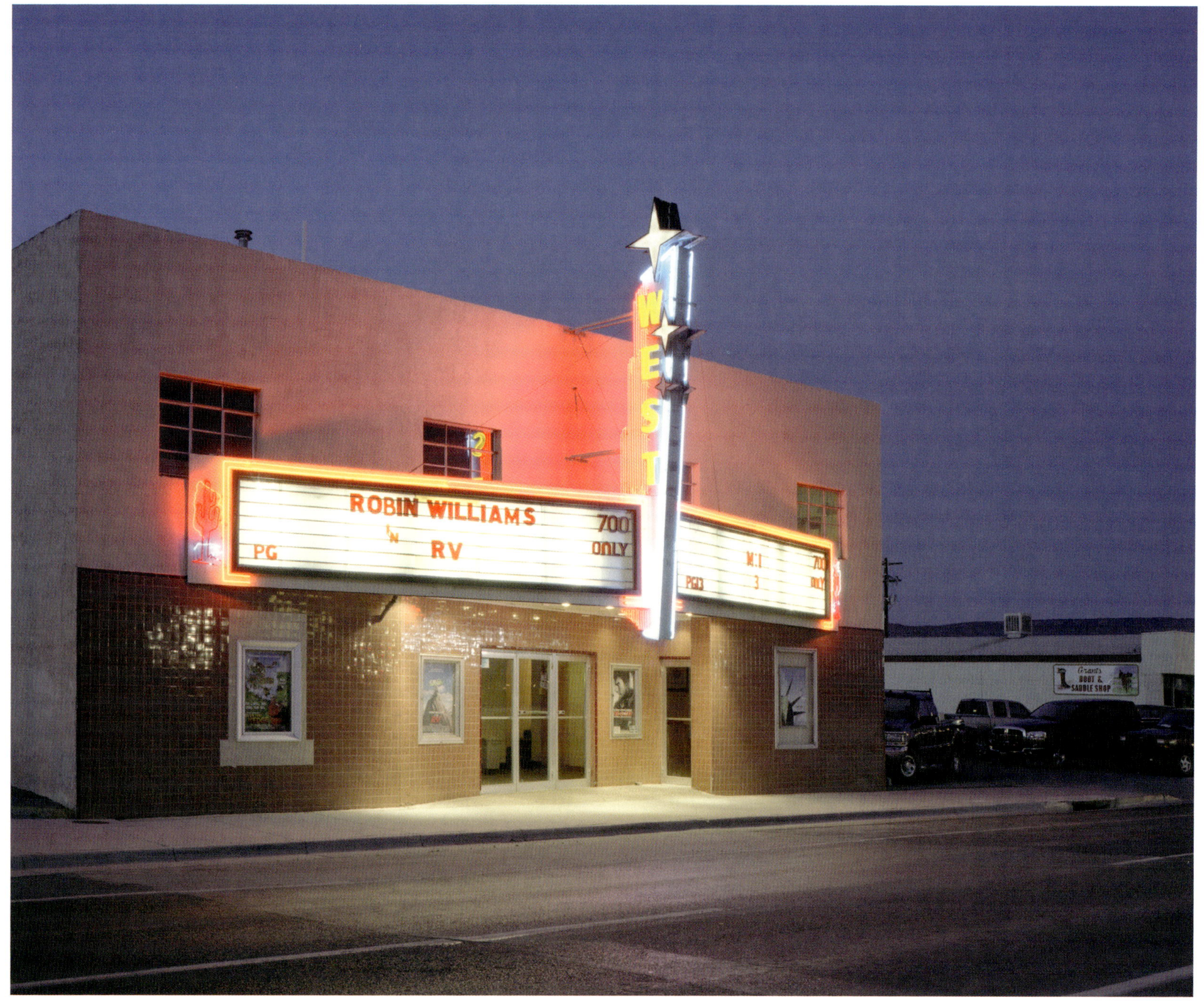

The West Theater, Grants, New Mexico
Opened in 1960 at the height of the golden years of neon

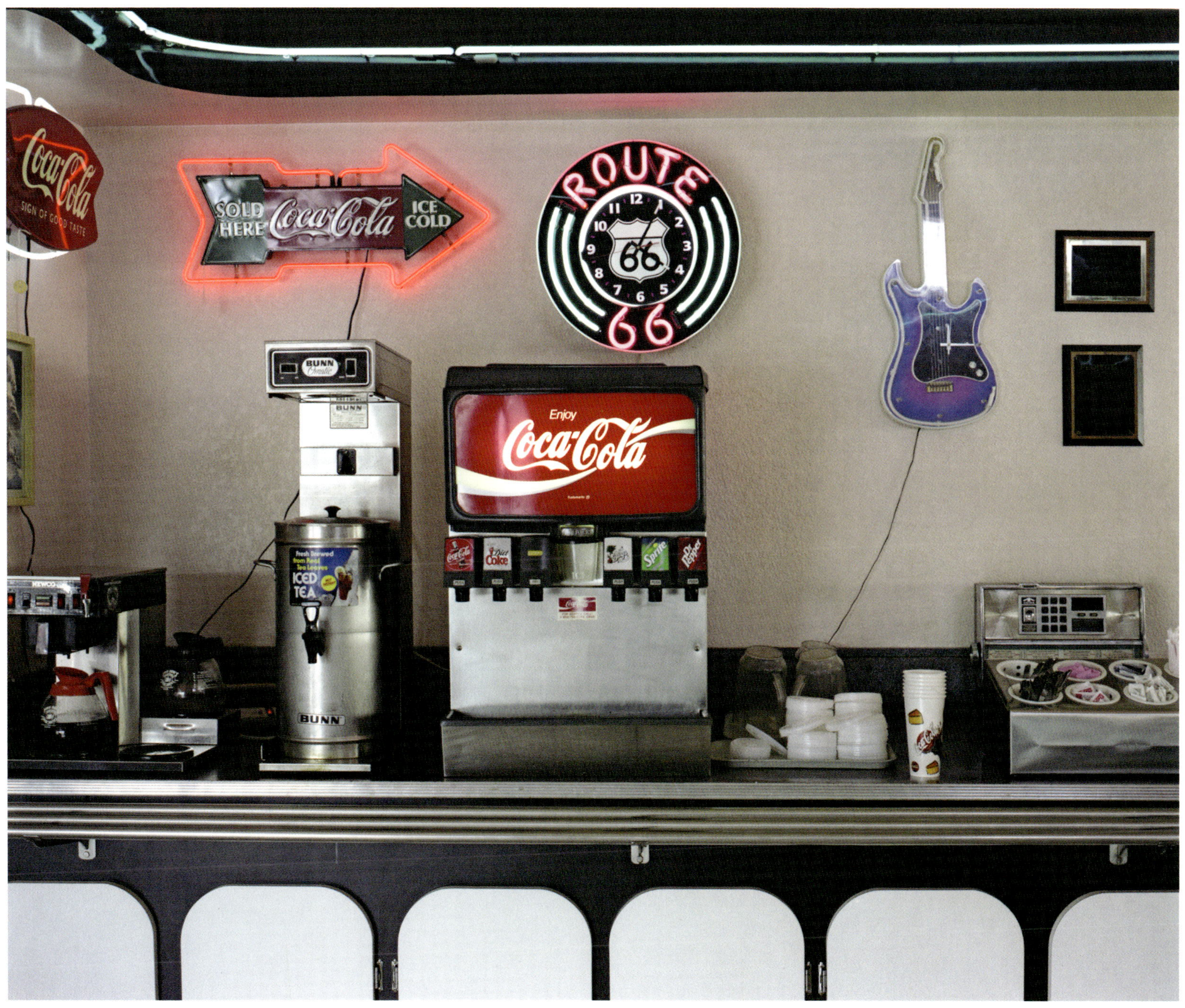

"Enjoy Coca Cola"

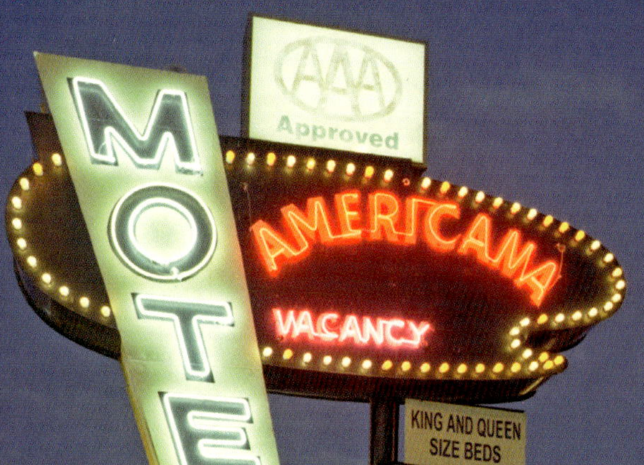
MOTEL
AAA
Approved
AMERICANA
VACANCY
KING AND QUEEN
SIZE BEDS
PHONES
REMOTE TV
HBO
AARP
21 95
ONE PERSON
Coca-Cola

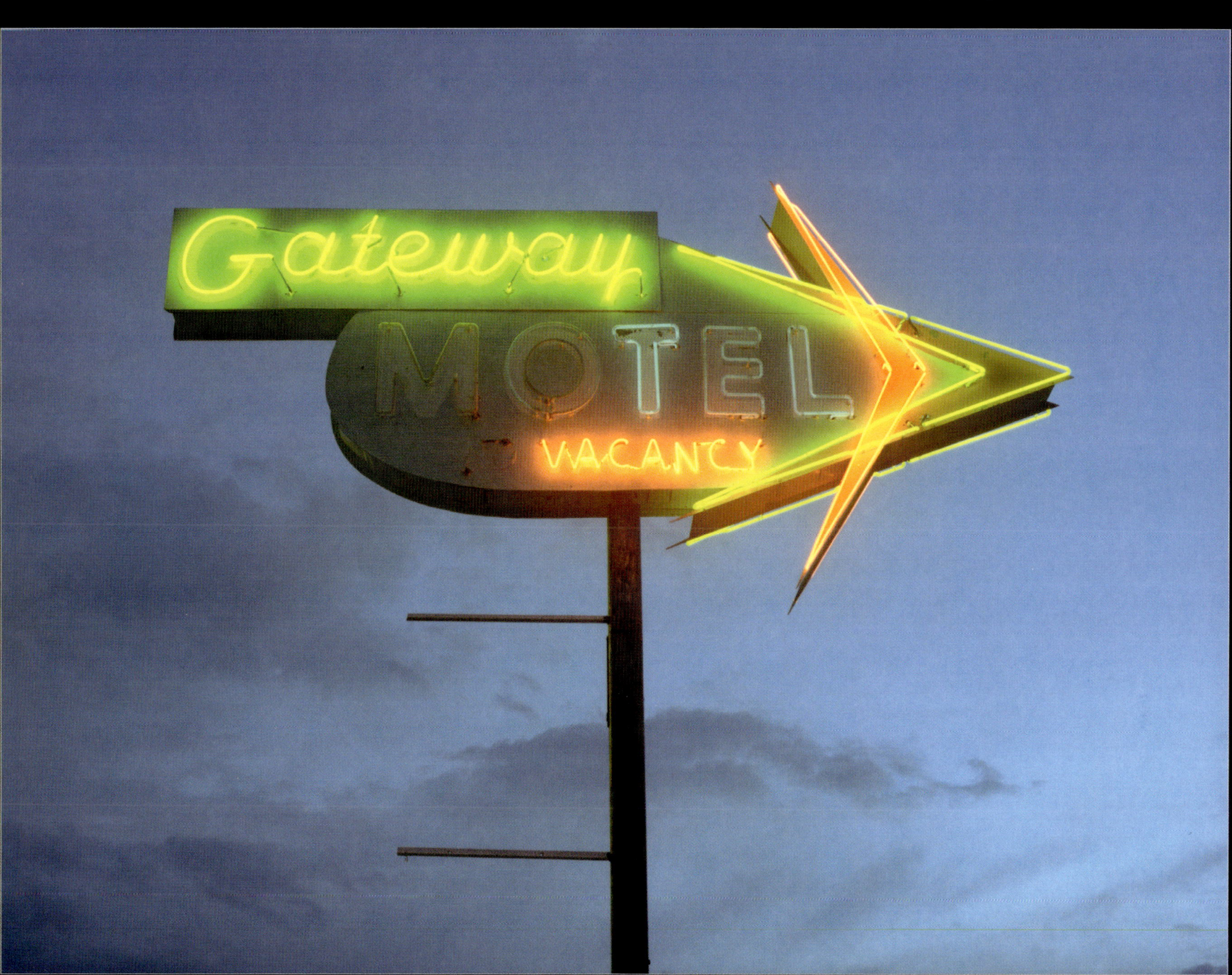

Gateway
MOTEL
VACANCY

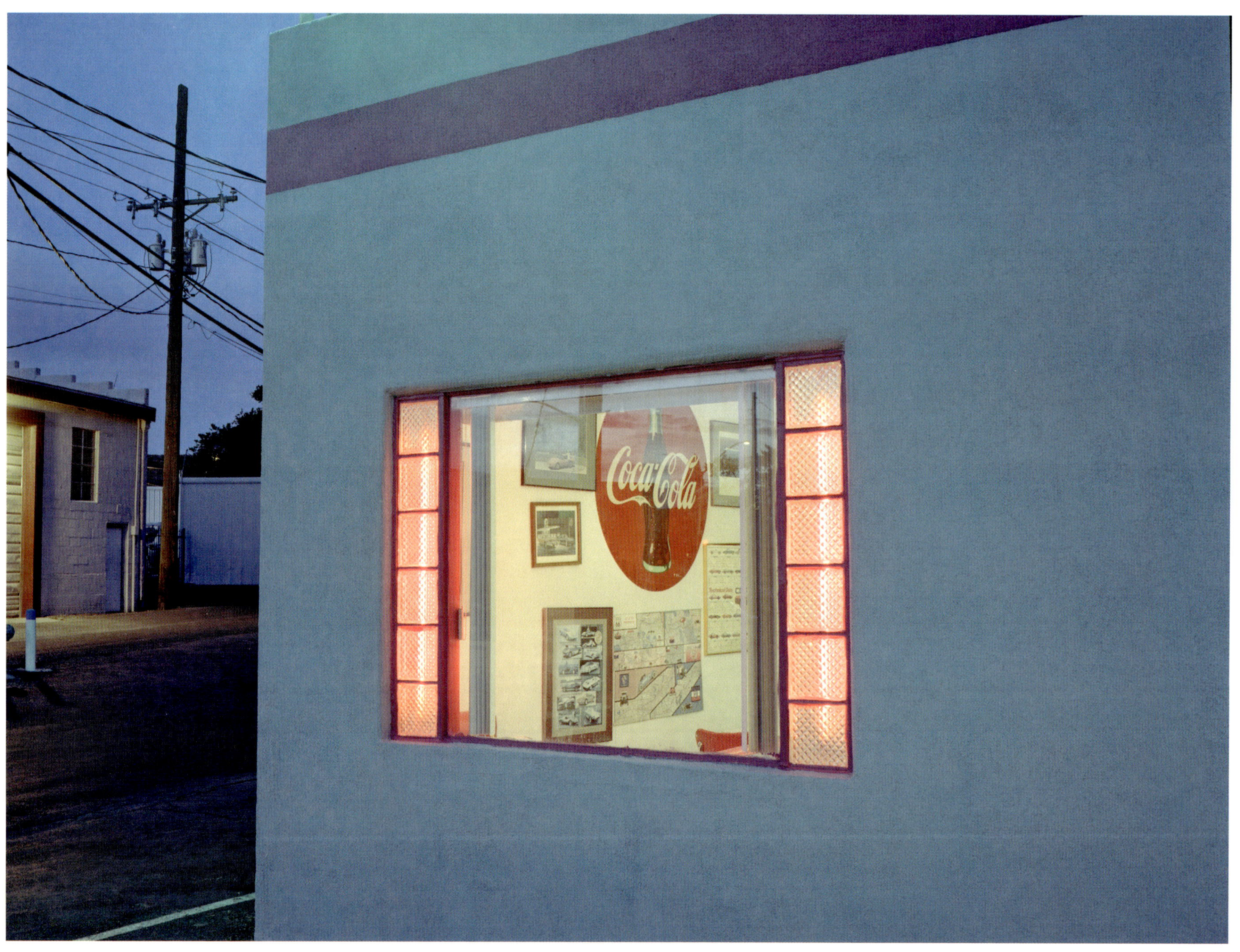

Mr D'z Diner, Kingman, Arizona

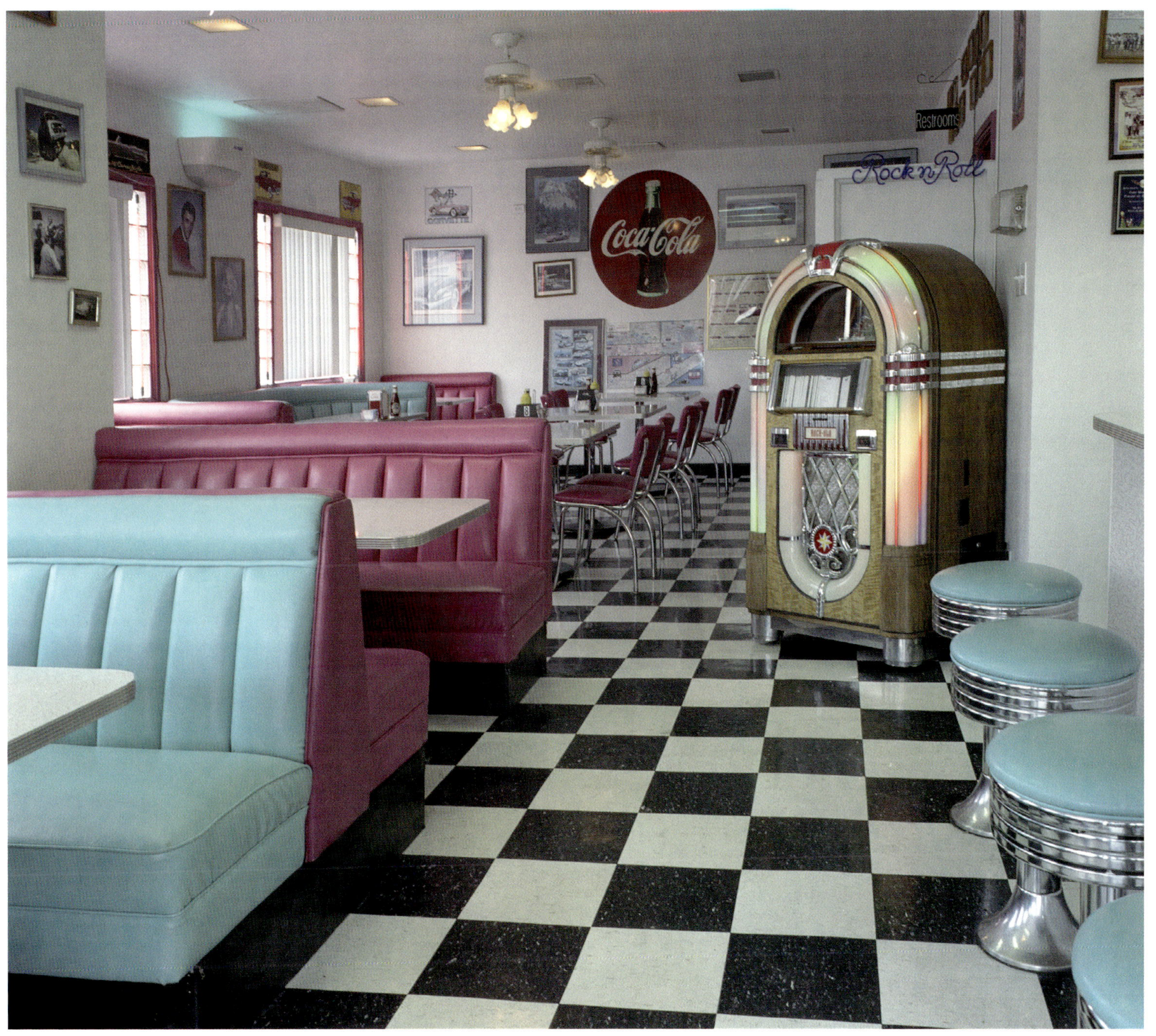

Interior of Mr D'z Diner

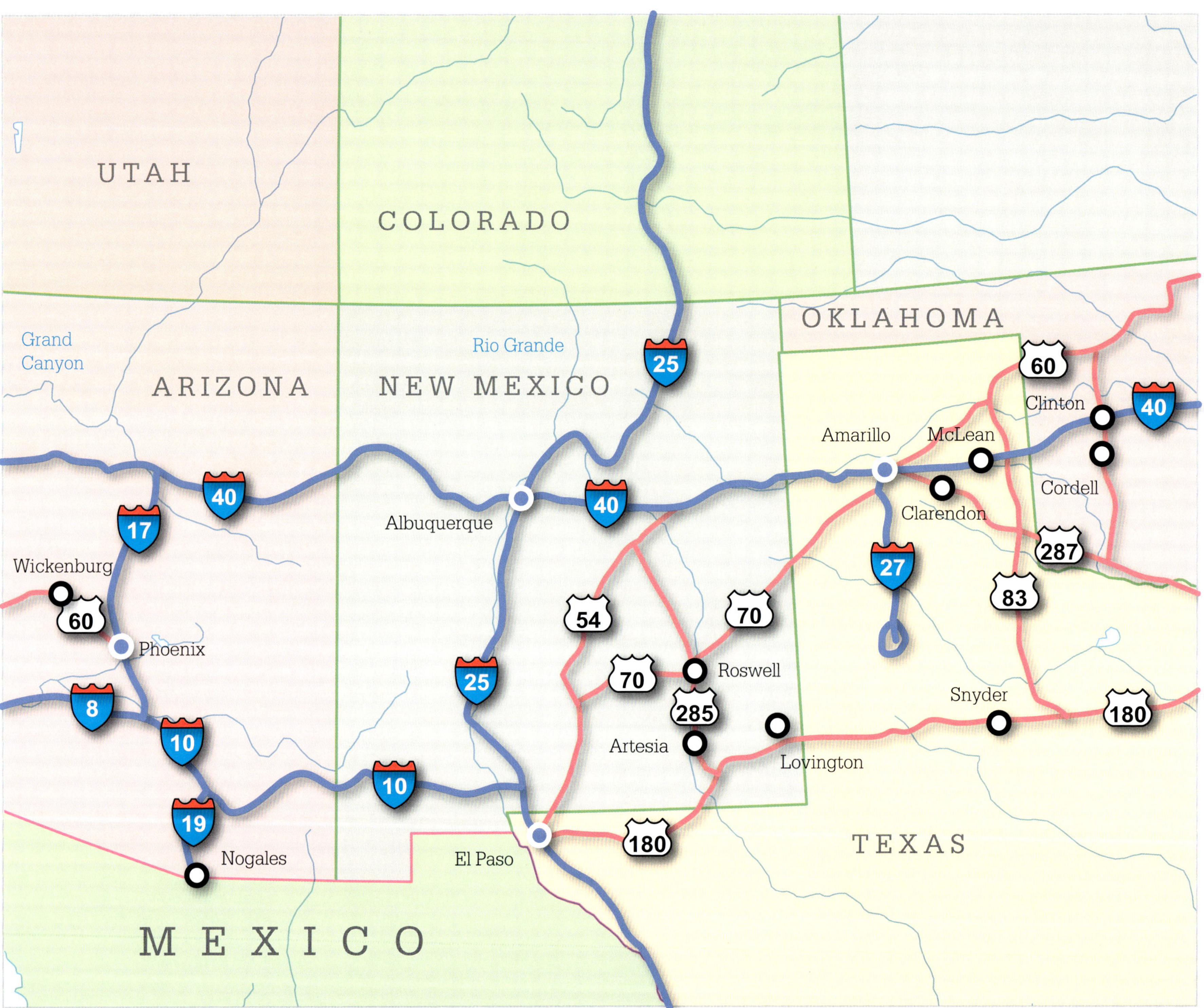
UTAH
COLORADO
OKLAHOMA
ARIZONA
NEW MEXICO
TEXAS
MEXICO
Grand Canyon
Rio Grande
Wickenburg
Phoenix
Nogales
Albuquerque
El Paso
Roswell
Artesia
Lovington
Amarillo
McLean
Clarendon
Clinton
Cordell
Snyder
25
40
17
40
8
10
19
10
25
54
70
70
285
180
27
60
40
60
287
83
180

The Southwestern Desert

Geographically and culturally, the southwestern states epitomize that sense of the open highway which is a lure for so many tourists. Here you find an expansive, colorful and varied landscape characterized by twisting canyons, large expanses of bone-dry desert, and clear blue skies. America has often been described as a melting pot, a description which applies well to this particular area. Culturally, it finds itself on the confluence of three distinct traditions, Hispanic, Anglo and Native American. This is also archetypal cowboy country, and characters wearing high-heeled boots and stetsons mingle comfortably with others who prefer to be in baseball caps.

The eastern fringe of the desert starts in central Arizona, in Wickenburg, a town that was founded on gold mining and ranching. The surrounding area is littered with the remains of ghost towns and redundant mines. Most old mining towns lie forgotten, but the occasional one has survived, often attracting a lively artist community and the concomitant cafe society.

Traveling south past the enormous urban sprawl of Phoenix, the fastest growing city within America and a magnet for retirees, the route heads across the Sonora Desert toward Tucson. Here it's uncomfortably warm even in April. It is easy to forget how southerly some of these Arizona cities are; sitting just 30 degrees north, Tucson lies on the same latitude as Marrakesh and Cairo. It once boasted some of the most flamboyant neon anywhere in America, particularly along the Miracle Mile, but this has all but disappeared.

Further south, toward Nogales and the Mexican border, the Hispanic influence is particularly strong. In common with many border towns, Nogales seems unsure of its identity, not typically American but at the same time certainly not Mexican. Go eastward and the landscape appears barren, a dry emptiness punctuated only by the occasional ghost town. Some old mining communities, such as Bisbee, have discovered a new lease of life, yet they still retain the wonderful old Victorian buildings that tell of a colorful past.

As you head into New Mexico, the terrain becomes increasingly dry and infertile; there is always the danger of encountering a severe dust storm. It is only as you approach the Rio Grande Valley that the landscape dramatically changes, the desert giving way to rich alluvial plains with neat farms and expansive fields of pecan trees. Between the Rio Grande and the Organ Mountains to the east is the city of Las Cruces, famed for its Spanish-style adobe buildings. As you travel north toward Socorro you encounter further desert scenery, not unlike that of the Australian Outback, with small communities thinly scattered around. Further east you find landscape that is much more agreeable, with the distant mountains always providing an interesting focal point. As the road continues to climb, the desert finally gives way to low-growing conifers.

Roswell, famed for its UFO connections, is situated among surroundings as desolate and inhospitable as you could care to imagine: limitless plains dotted with cactus. If aliens truly wanted to land on earth undetected, this would indeed be the ideal spot. Artesia, due south of Roswell, is visually an attractive town, the attraction being tempered more than somewhat by the omnipresent smell from the nearby petrochemical complex. If you go east from Artesia in the direction of Lovington and finally toward Texas, you reach the desolate plains of the Llano Estacado, where thousands upon thousands of nodding donkeys fill the scene as far as the eye can see.

This terrain continues over the border and, although the fertility

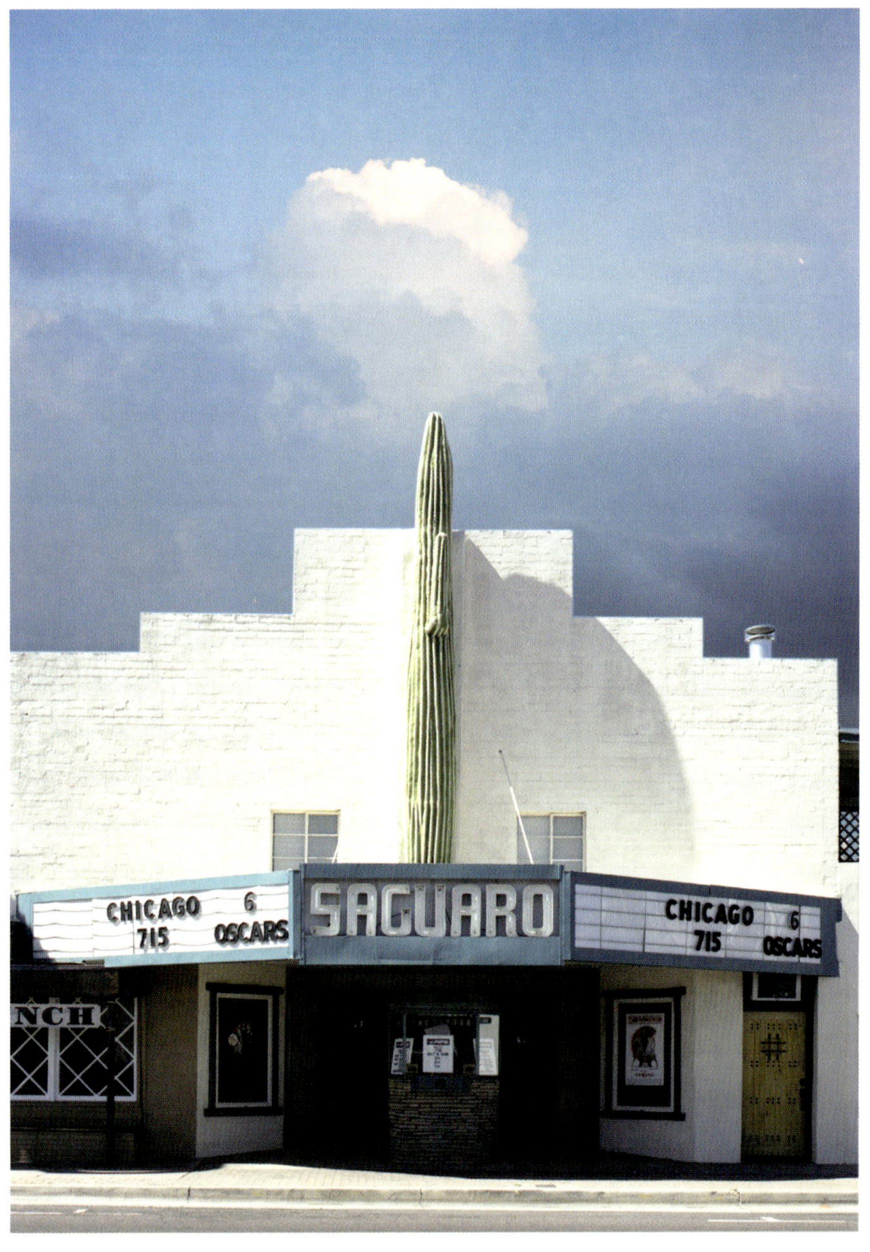

The Saguaro Theater, Wickenburg, Arizona
Opened in 1948, it now hosts both movies and live entertainment

of the land markedly improves, the nature of the geography is unambiguously spelled out in place-names like Brownfields and Plains. In this western part of Texas, the scenery remains resolutely flat, with sprawling plowed fields of cotton interspersed by oil derricks and scrub.

Traveling northward toward the panhandle, the road steadily climbs, the countryside becomes more undulating, and small red canyons begin to appear. Intensive agriculture gives way to ranching, and the population slowly thins out. Here and there, small clusters of cattle group together round a convenient waterhole. The towns in this part of Texas are often quite charming, downtown being built round a handsome square with the cinema as focal point. Such scenes suggest an era when the entire population flooded into the center on a Saturday evening intent on having a great time. McLean, in the north of the state and situated on Historic Route 66, was once a major stopping-off point, but in common with so many communities along this famous old highway has been slipping into decline. McLean was once famed for its dazzling neon lighting, but with diminishing trade this has all but disappeared.

Journeying through this southwestern pocket of America is truly a romantic experience. The route takes you through some of the most breathtaking scenery America has to offer. Fans of the cult movie *Thelma & Louise* (1991) certainly won't be disappointed by the byroads of this gloriously enchanting region!

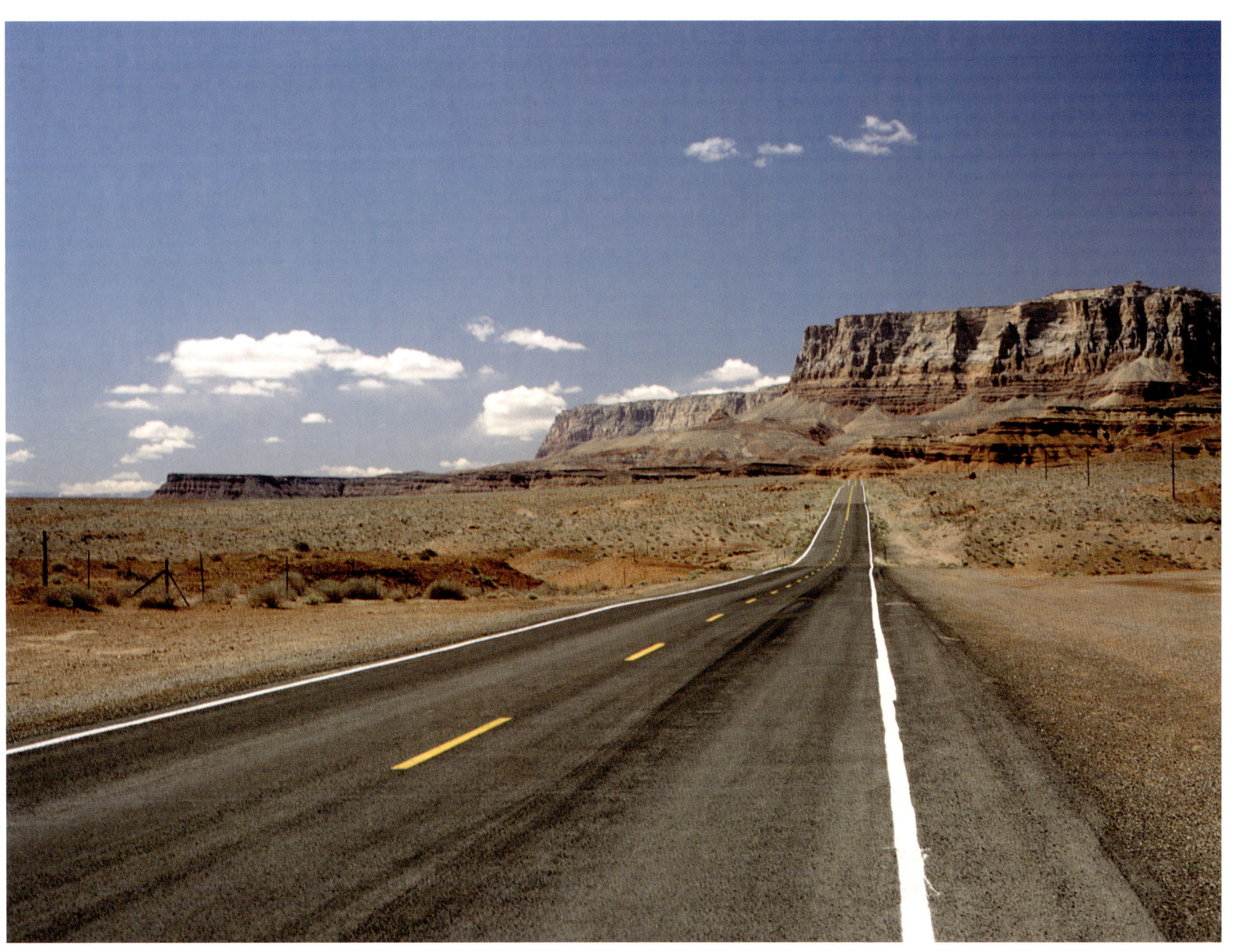

Highway in southern Arizona, a landscape characterised by twisting canyons, large expanses of bone-dry desert and clear blue skies

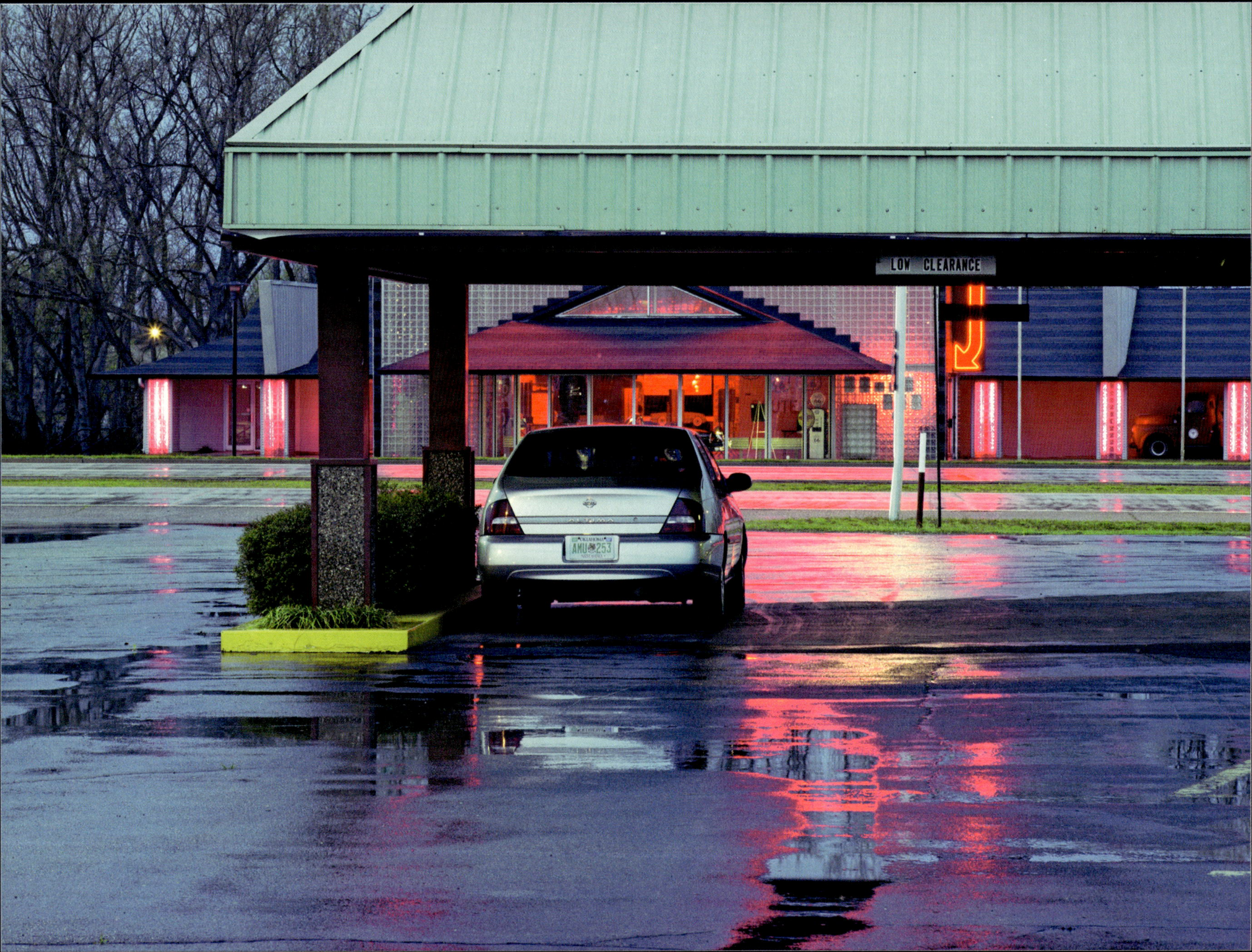

LOW CLEARANCE

AMU 253

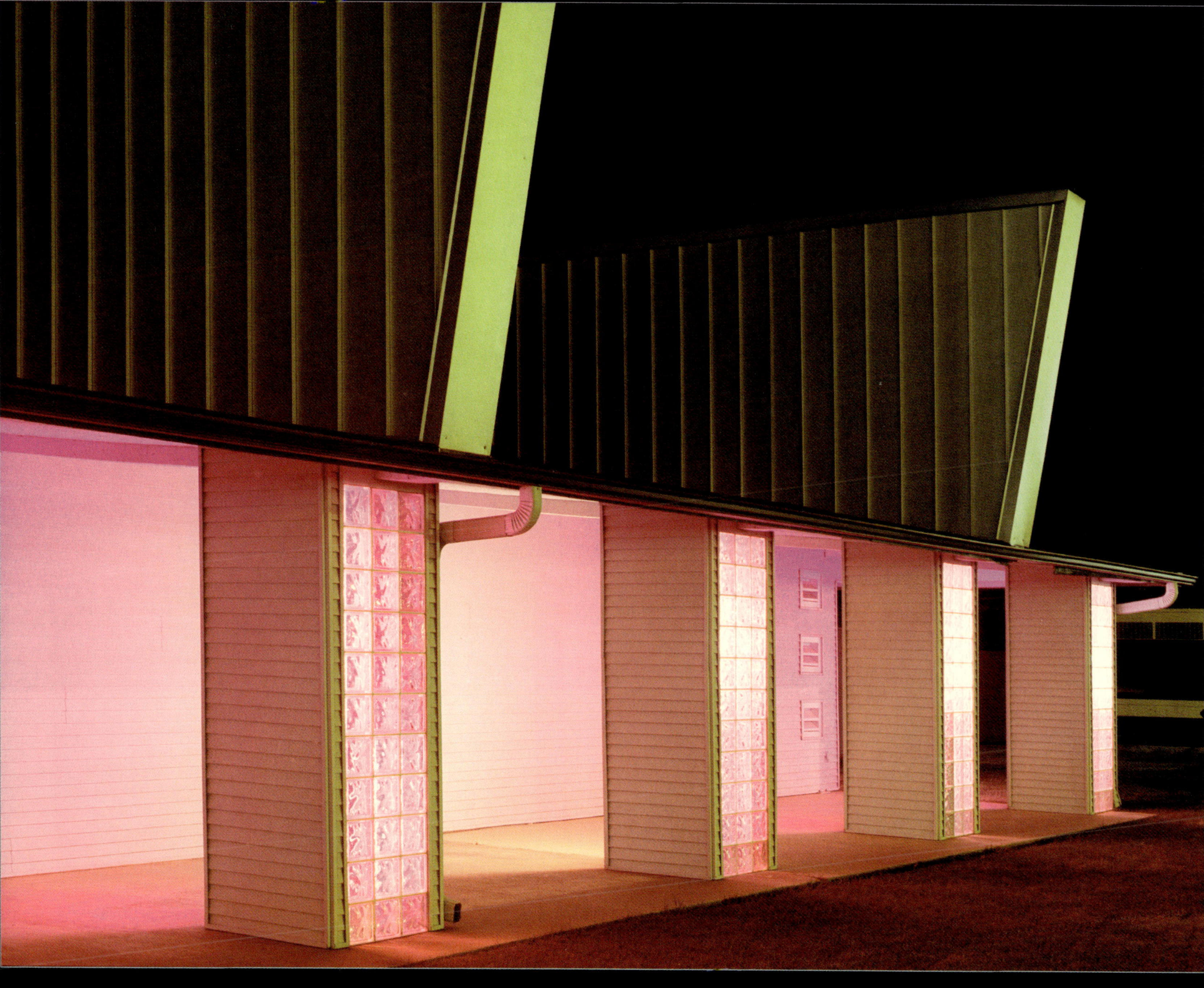

Museum, Clinton, Oklahoma

An excellent museum that showcases the history of the Mother Road

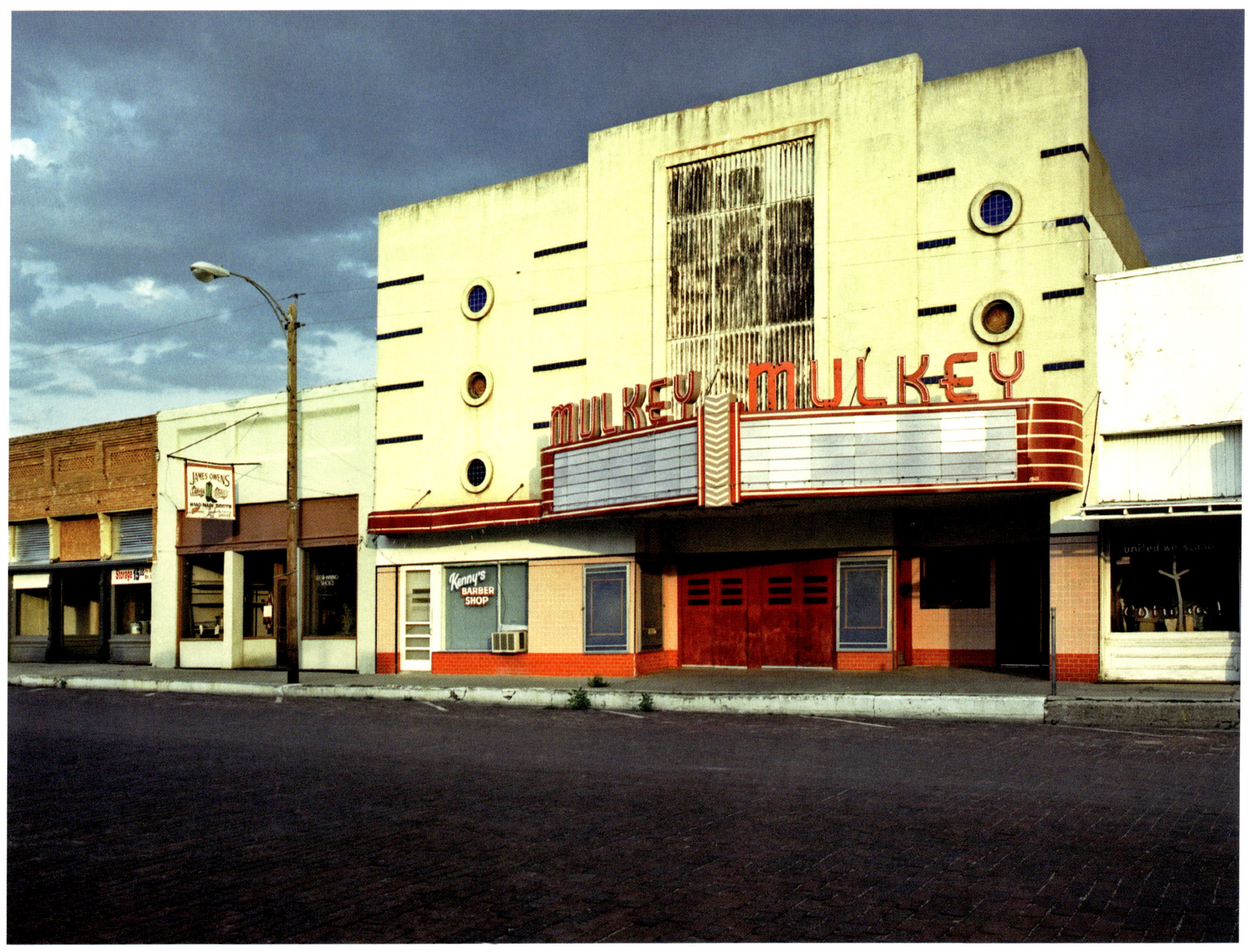

Mulkey Theatre, Clarendon, Texas

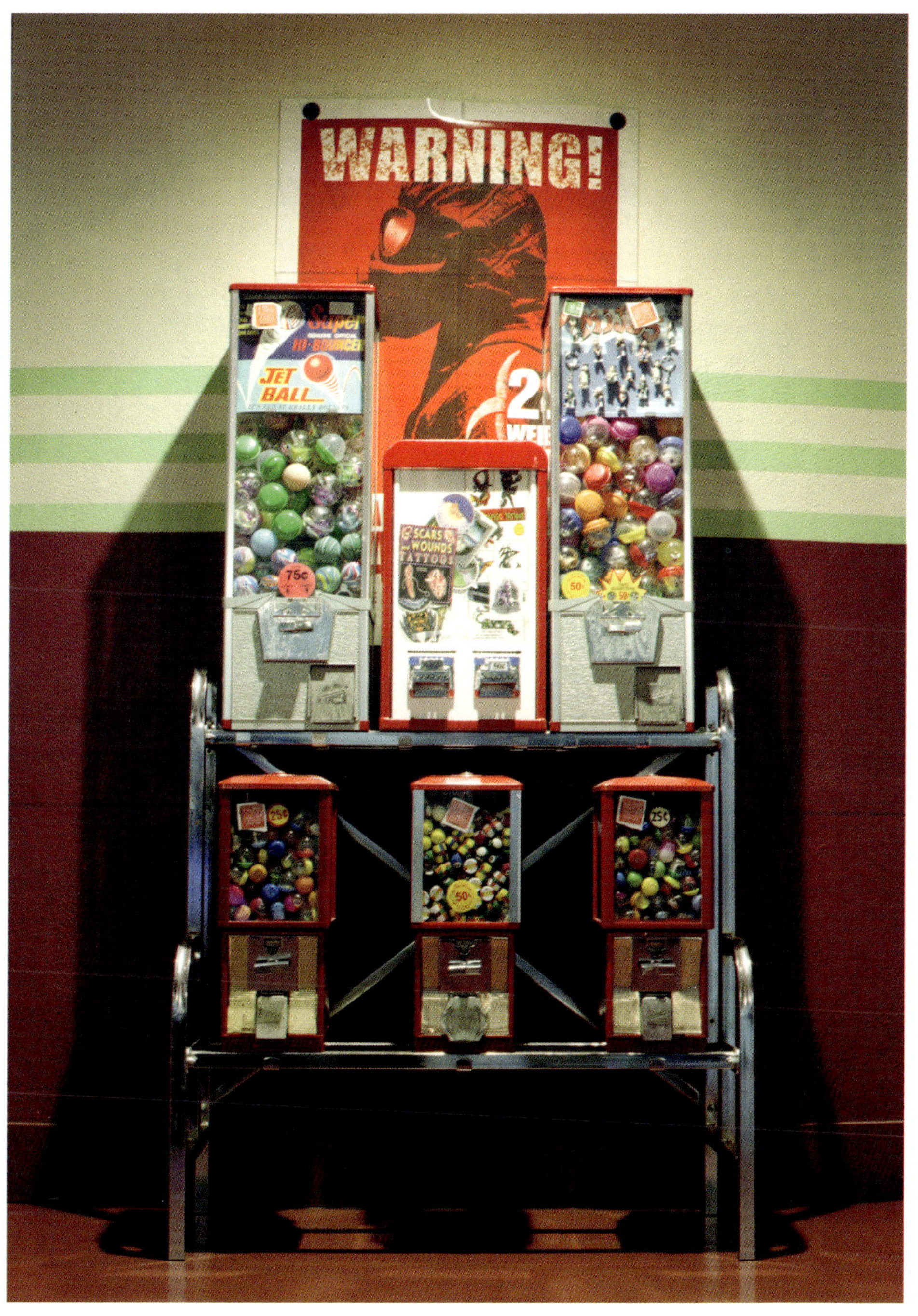

Bubble-gum machine, Washita Theatre, Cordell, Oklahoma

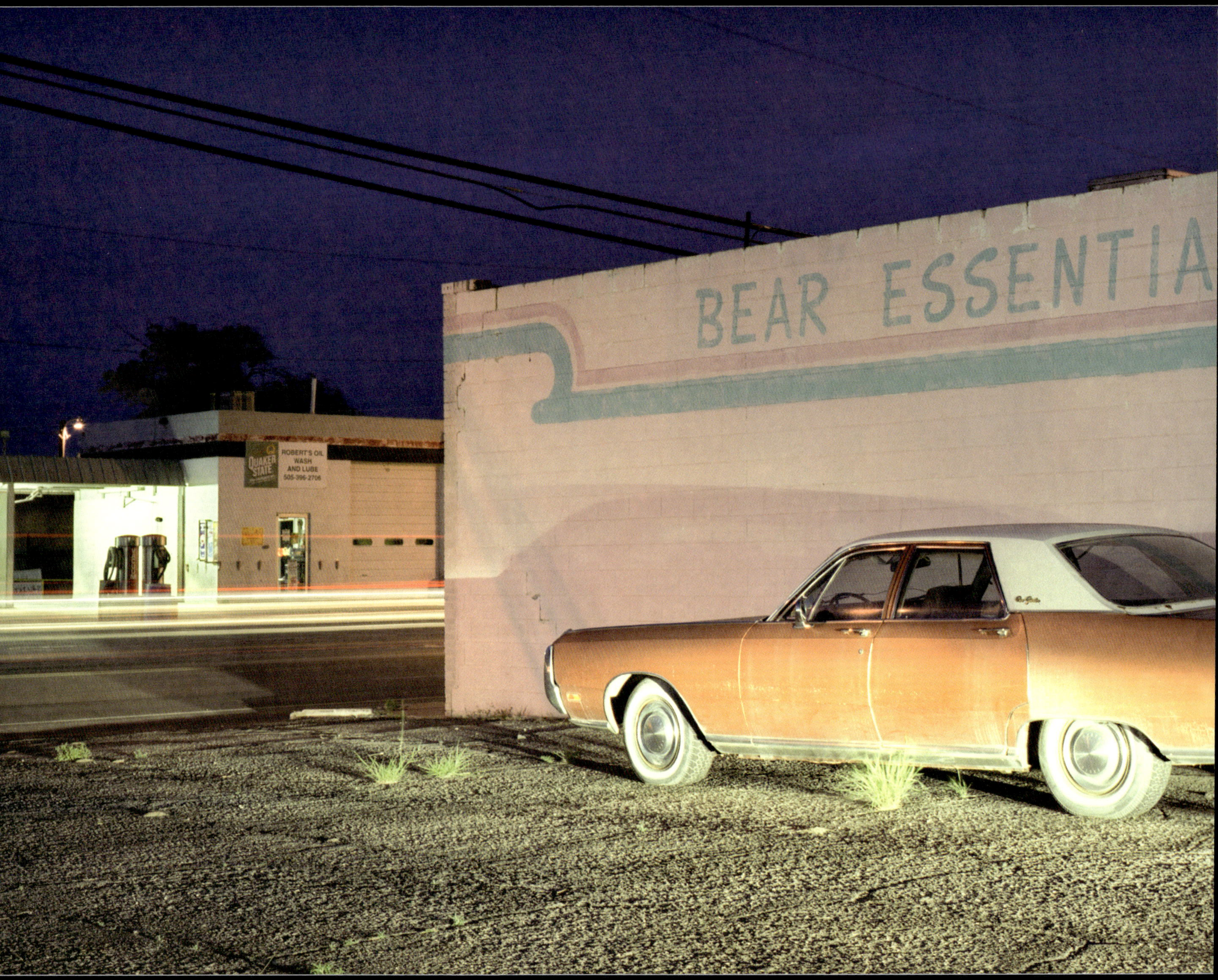
BEAR ESSENTIA
QUAKER
STATE
ROBERT'S OIL
WASH
AND LUBE
505-396-2706

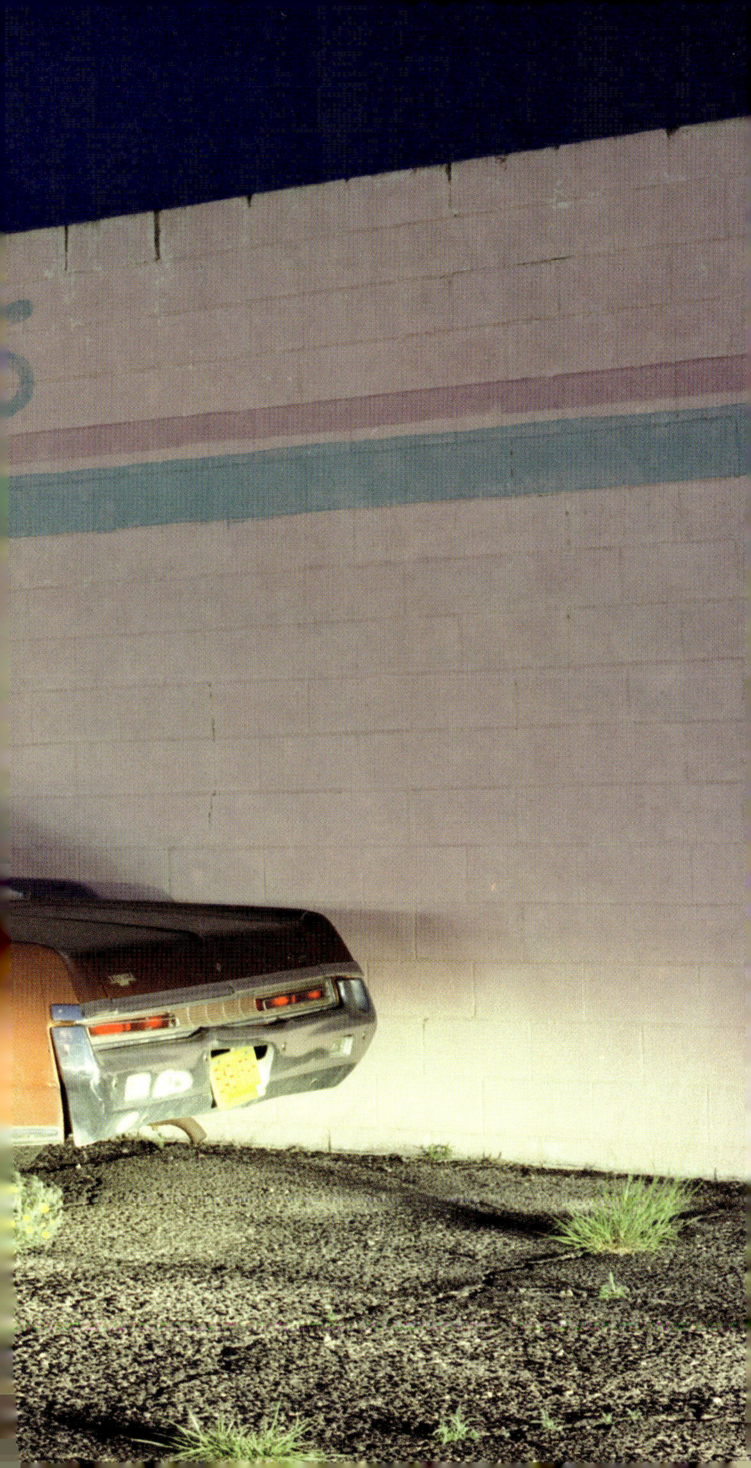

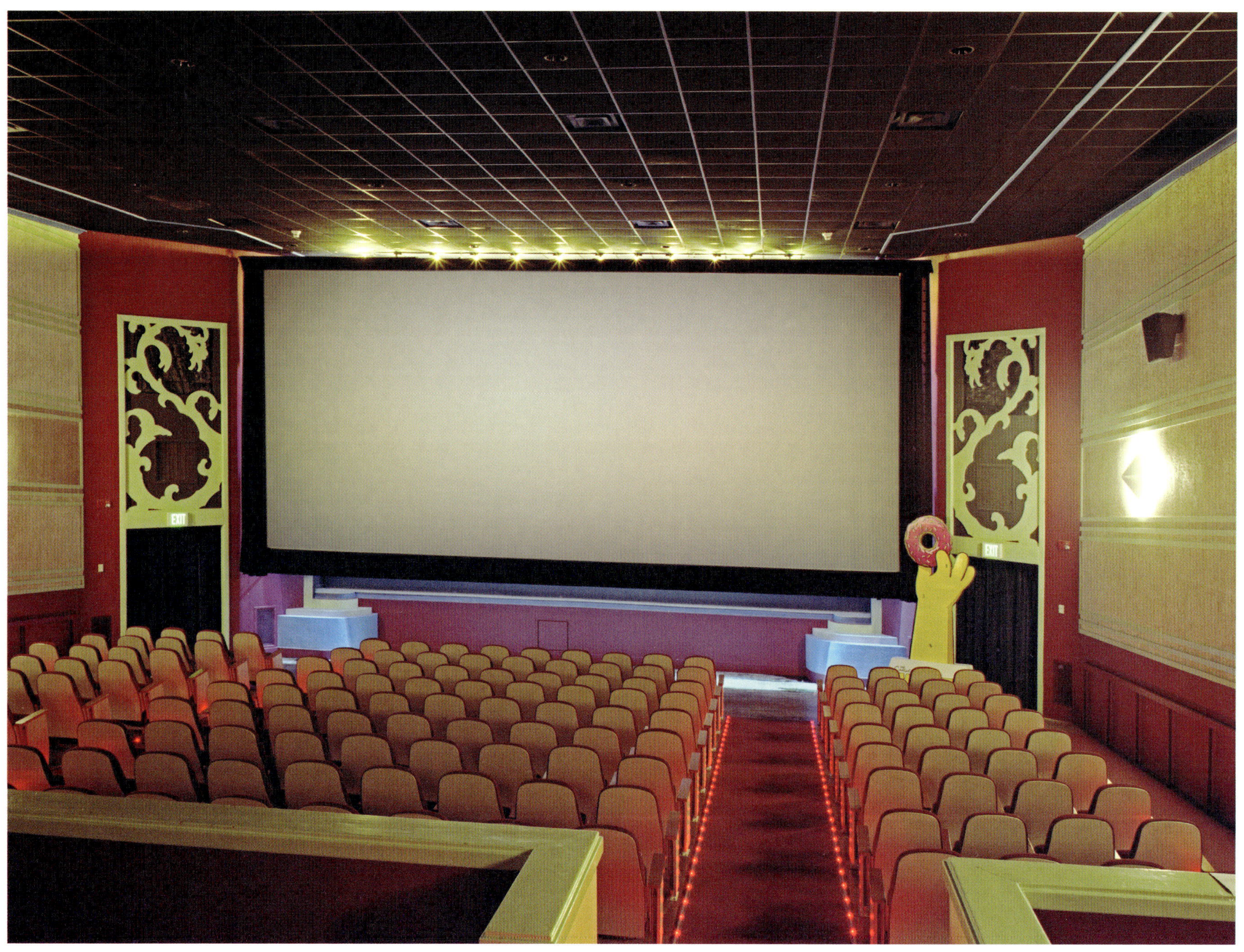

The auditorium, The Washita Theatre, Cordell, Oklahoma

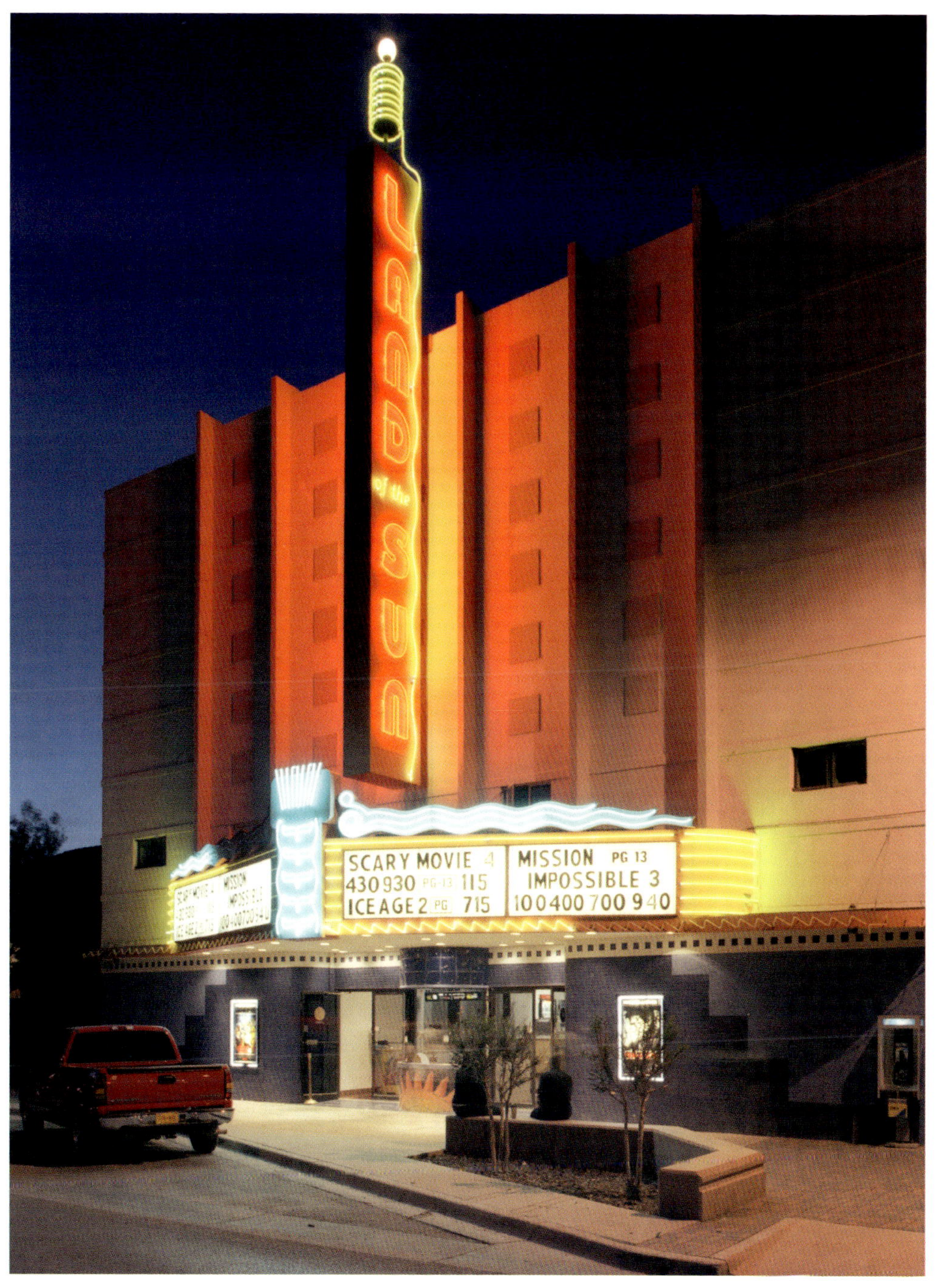

"Land of the Sun", Artesia, New Mexico
The neon is a replica of the original 1947 sign

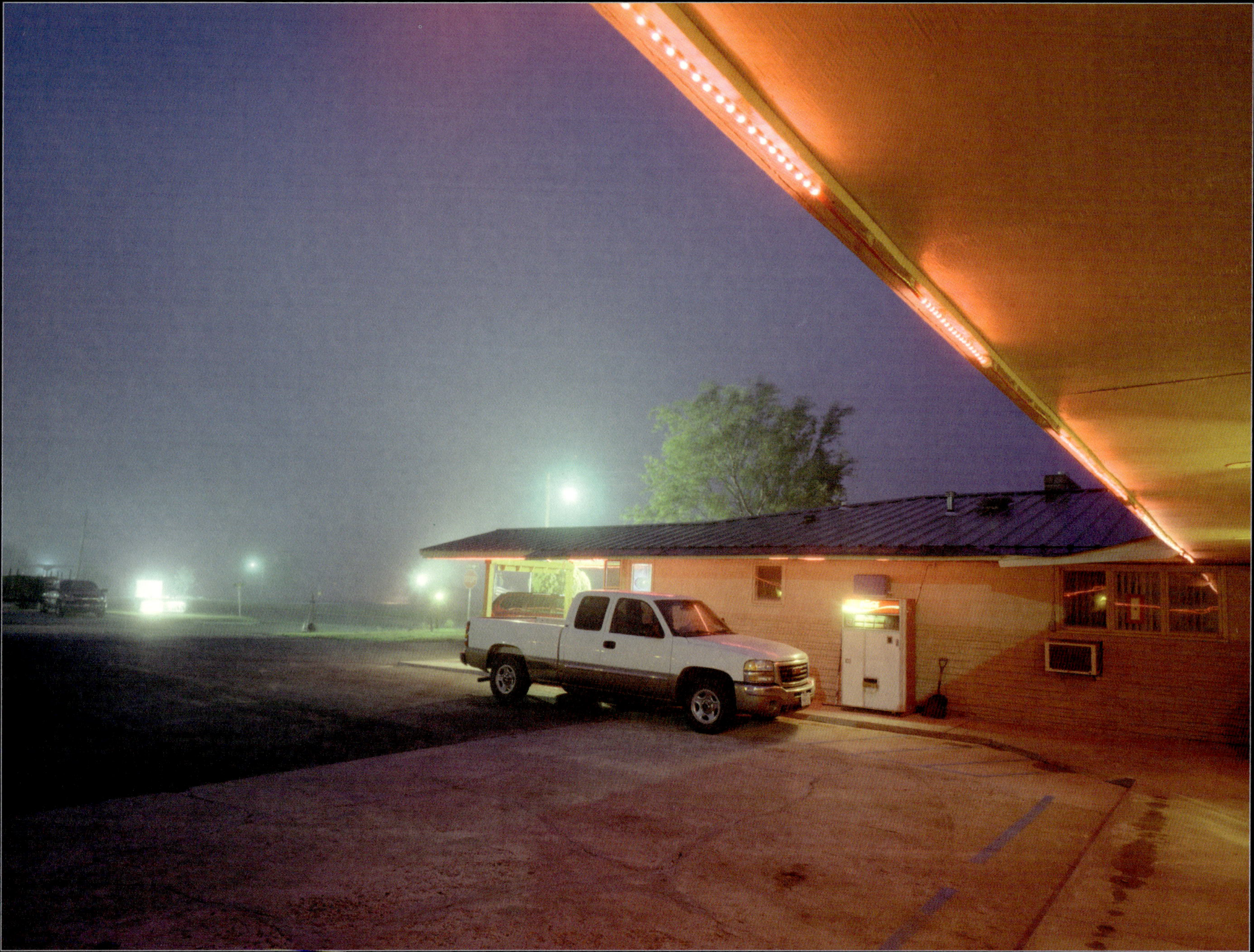

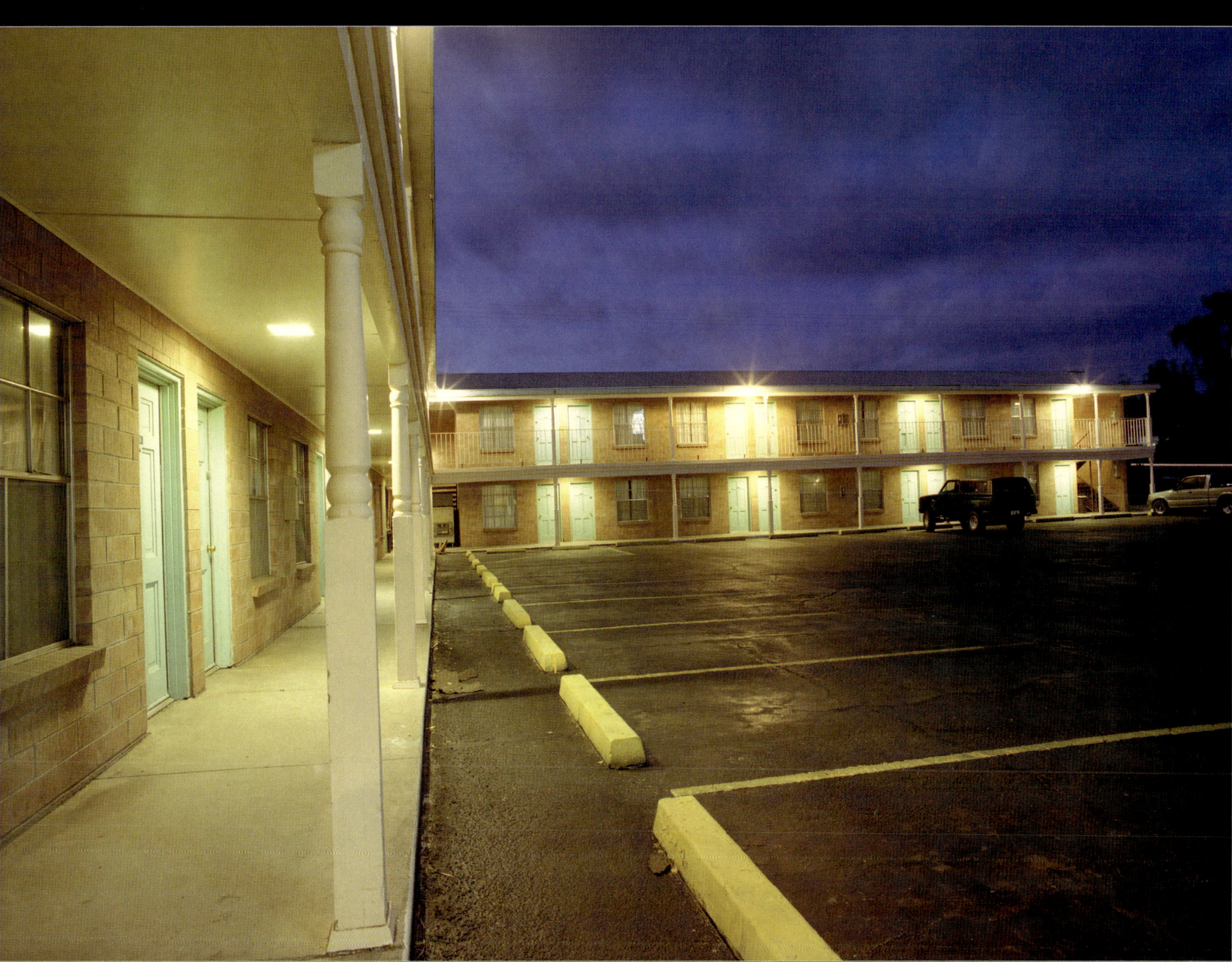

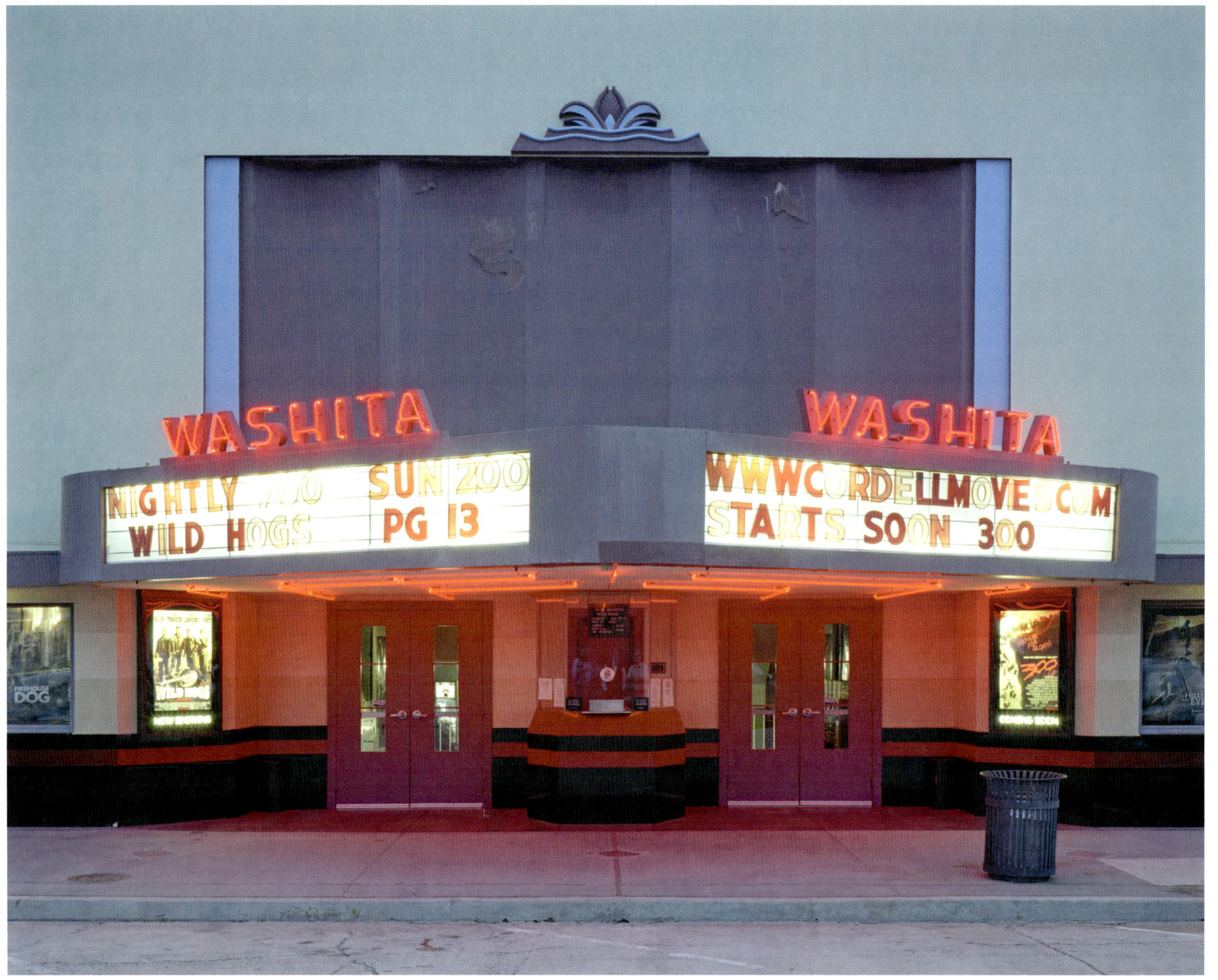

The Washita, Cordell, Oklahoma

Built in the 1940s in Art Deco style, the theatre was extensively refurbished in 2000 at a cost of over $1,000,000

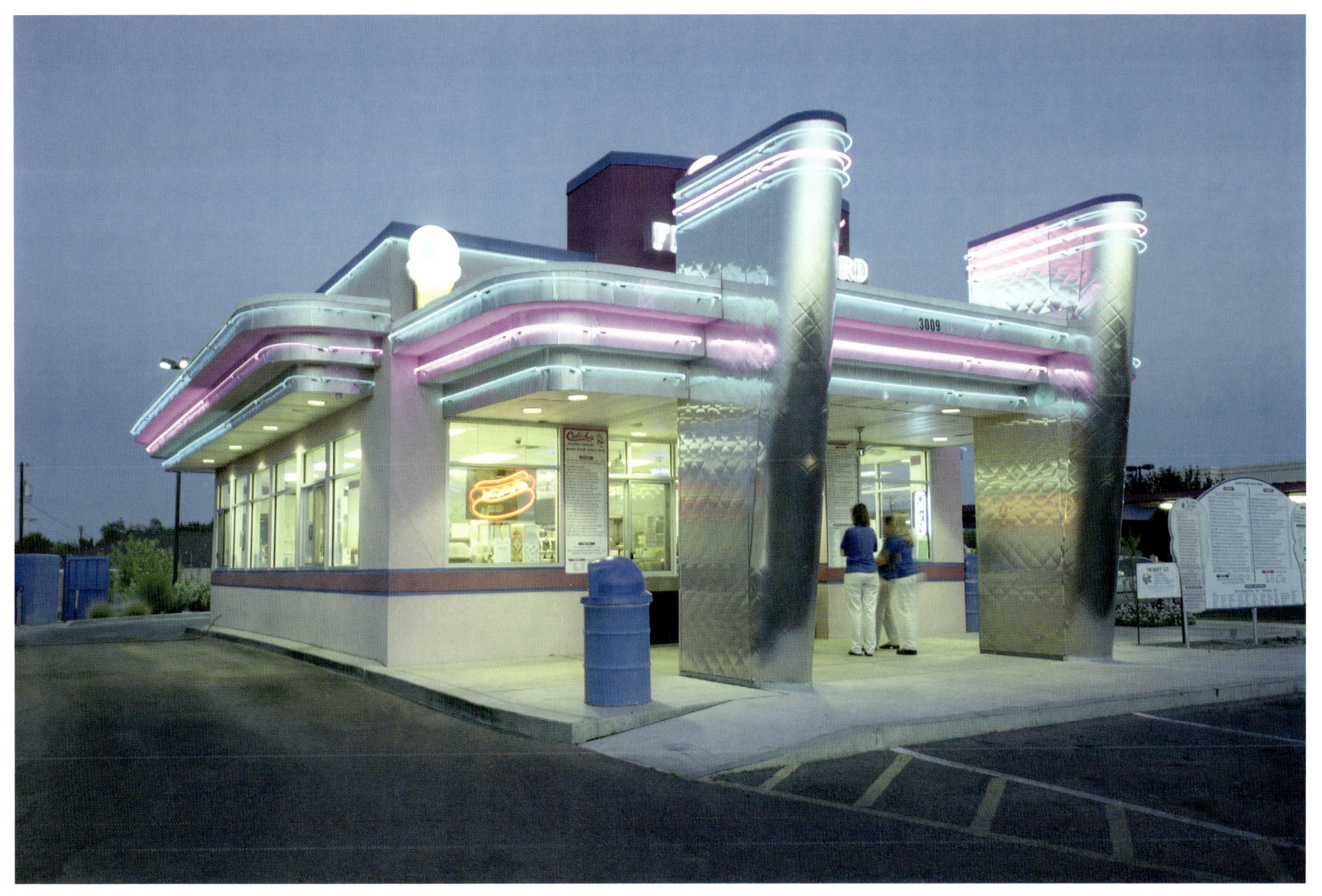

Ice-cream parlour, Roswell, New Mexico

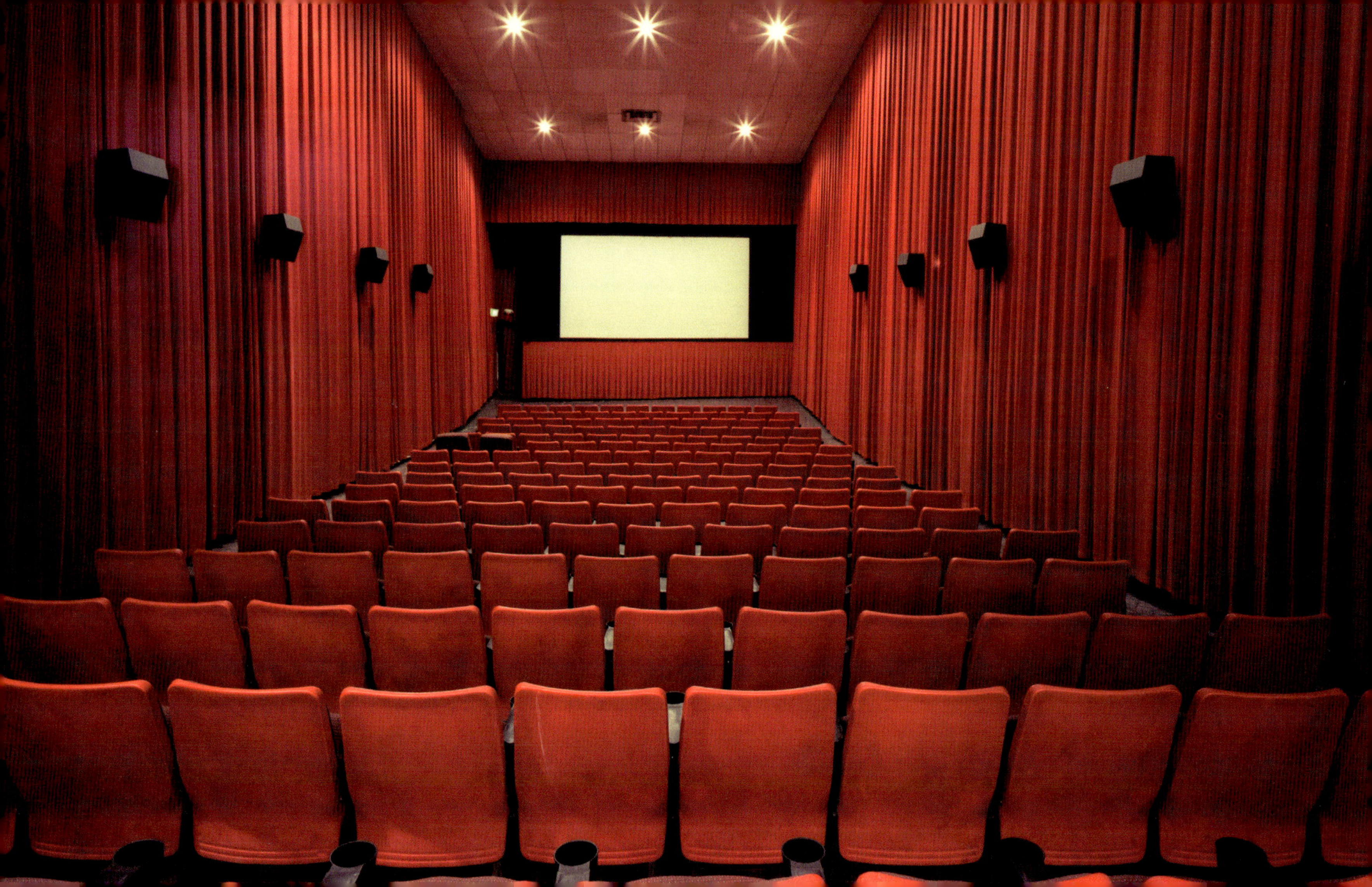

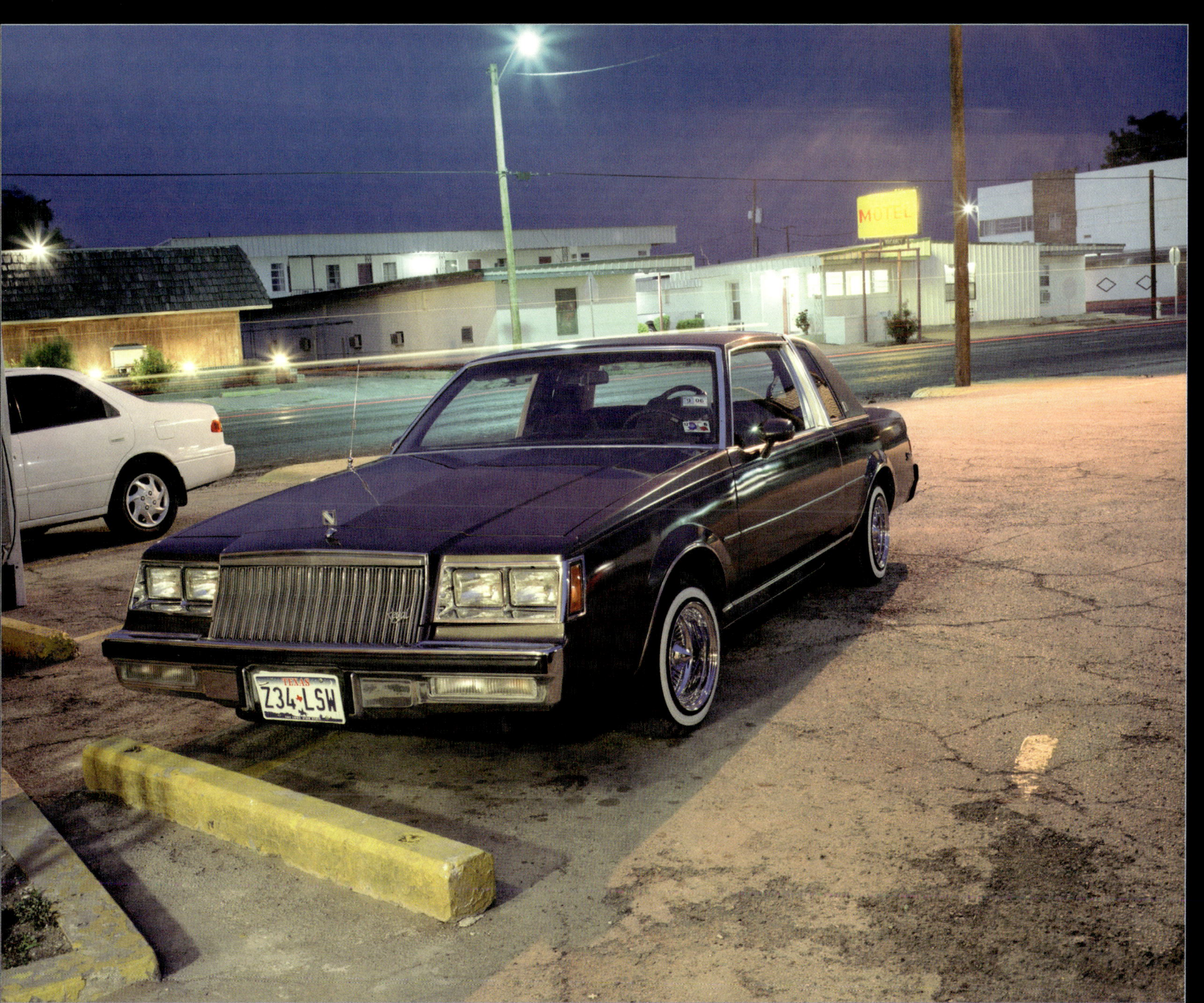
TEXAS
Z34•LSW
MOTEL

6 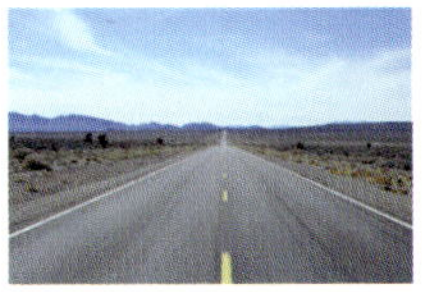
Tony Worobiec
Mamiya 7
43 mm
f16, 1/125th
Fuji Reala 100

7 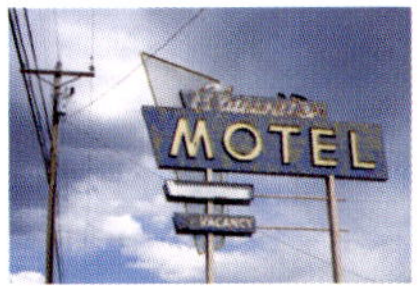
Eva Worobiec
Mamiya 645
80 mm
f16, 1/60th
Fuji Reala 100

8 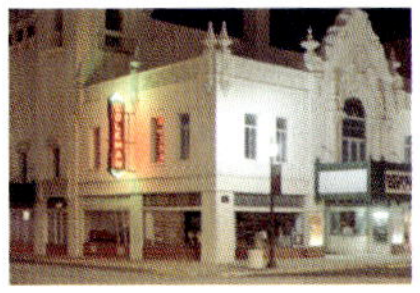
Eva Worobiec
Mamiya 645
80 mm
f16, 1 min
Kodak Portra 160

9 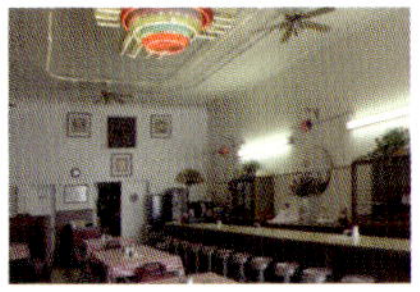
Eva Worobiec
Mamiya 645
50 mm
f22, 2 sec
Kodak Portra 400

9 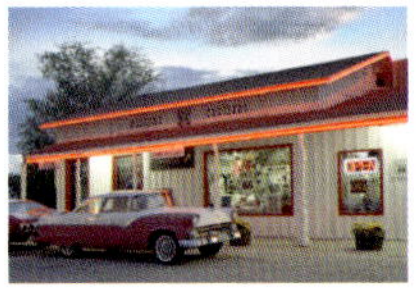
Tony Worobiec
Pentax 67
45 mm
f16, 2 sec
Fuji Reala 100

11 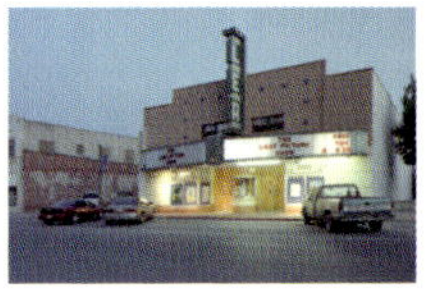
Eva Worobiec
Mamiya 645
50 mm
f16, 30 sec
NPH 400

14 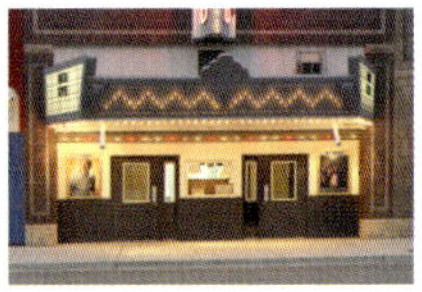
Eva Worobiec
Mamiya 645
50 mm
f16, 15 sec
Fuji Reala 100

15 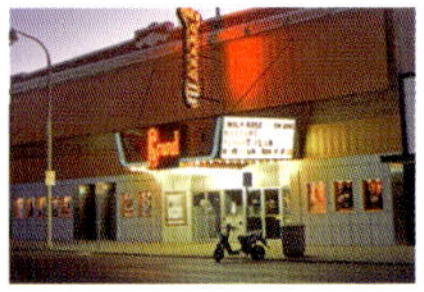
Eva Worobiec
Mamiya 645
45 mm
f16, 30 sec
Fuji Reala 100

16 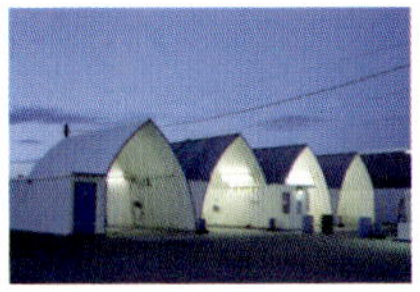
Tony Worobiec
Pentax 67
45 mm
f11, 2 min
Fuji Reala 100

17 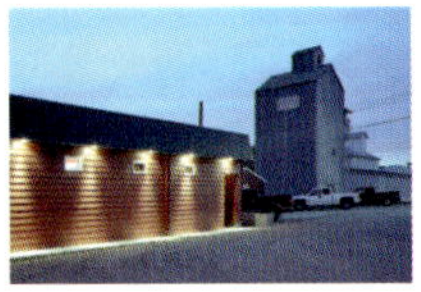
Tony Worobiec
Mamiya 7
43 mm
f16, 1 min
Fuji Reala 100

18 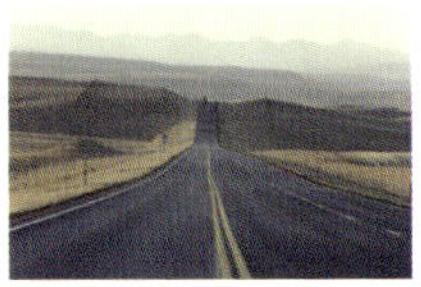
Tony Worobiec
Mamiya C330
250 mm
f45, 1 sec
Kodak Portra 400

19 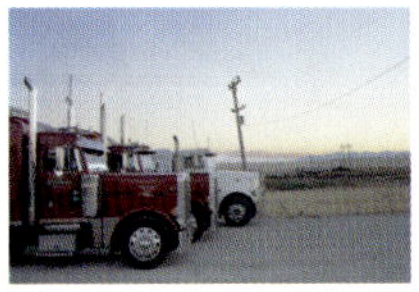
Tony Worobiec
Pentax 67
90 mm
f22, 30 sec
Fuji Reala 100

20 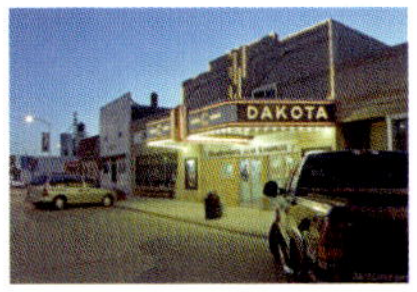
Eva Worobiec
Fuji 69
65 mm
f16, 8 sec
Kodak Portra 400

21 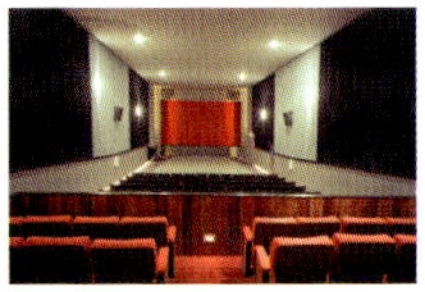
Tony Worobiec
Mamiya 7
43 mm
f16, 2 min
Fuji Reala 100

22 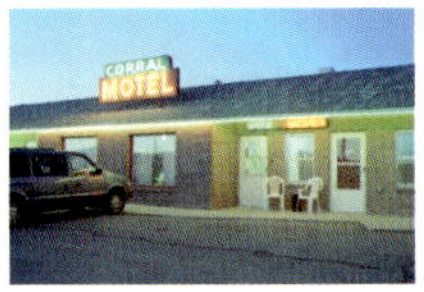
Tony Worobiec
Pentax 67
90 mm
f22, 2 min
Fuji Reala 100

23 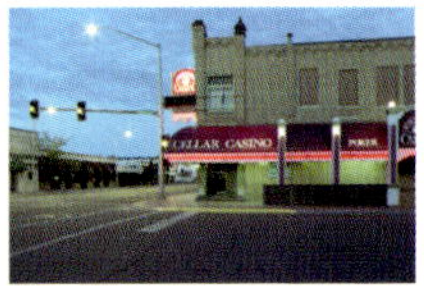
Tony Worobiec
Pentax 67
90 mm
f22, 1 min
Fuji Reala 100

24 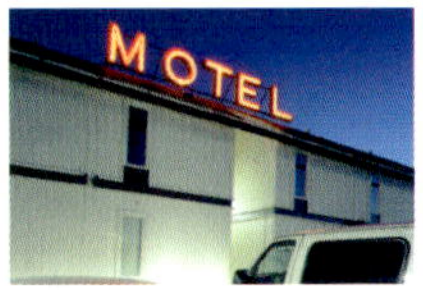
Tony Worobiec
Mamiya 7
43 mm
f16, 3 min
Fuji Reala 100

25 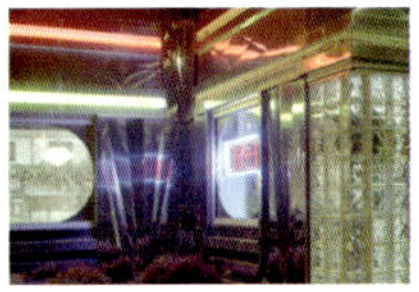
Eva Worobiec
Mamiya 645
80 mm
f16, 1 min
Kodak Portra 160

26 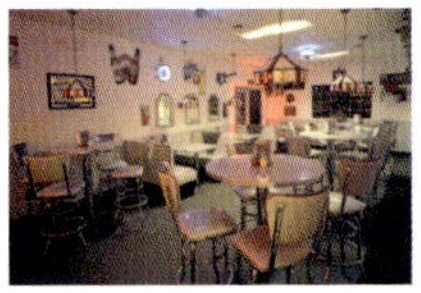
Tony Worobiec
Mamiya 7
43 mm
f16, 30 sec
Kodak Portra 400

27 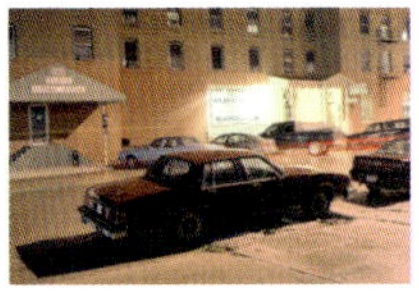
Tony Worobiec
Pentax 67
45 mm
f11, 3 min
Fuji Reala 100

28 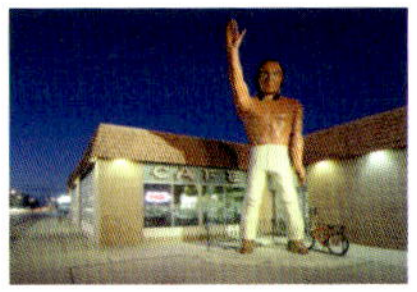
Tony Worobiec
Pentax 67
45 mm
f16, 1 min
Fuji Reala 100

29 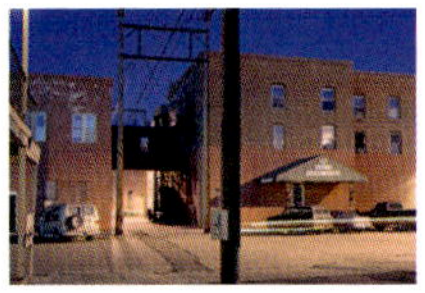
Tony Worobiec
Pentax 67
45 mm
f16, 5 min
Fuji Reala 100

30 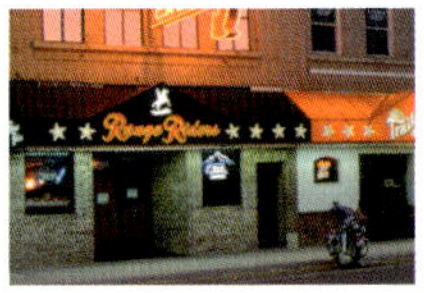
Eva Worobiec
Mamiya 645
50 mm
f16, 4 sec
NPH 400

31 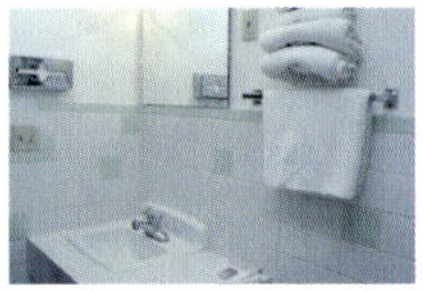
Tony Worobiec
Mamiya 7
43 mm
f16, 1 sec, flash
Kodak Portra 400

32/33 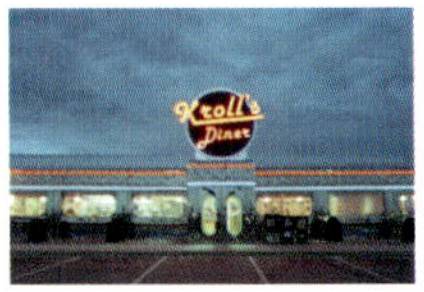
Eva Worobiec
Mamiya 645
45 mm
f16, 8 sec
Fuji Reala 100

34 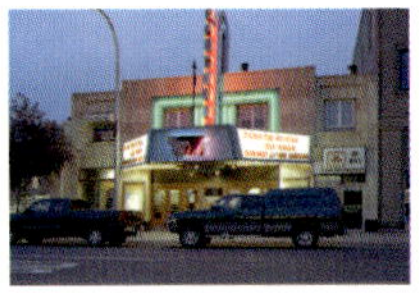
Tony Worobiec
Pentax 67
90 mm
f16, 1 min
Fuji Reala 100

35 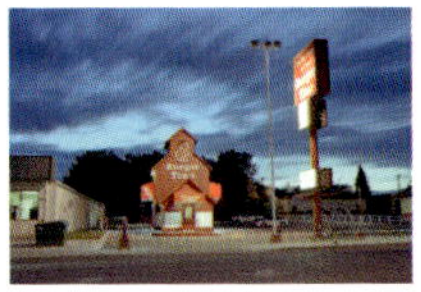
Tony Worobiec
Mamiya 7
80 mm
f16, 1 min
Fuji Reala 100

38 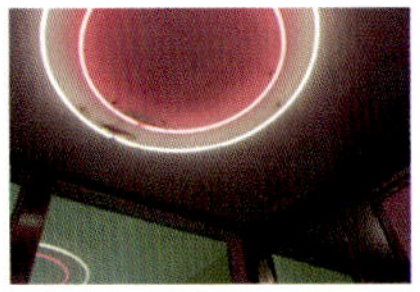
Tony Worobiec
Mamiya 7
43 mm
f11, 4 sec
Fuji Reala 100

39 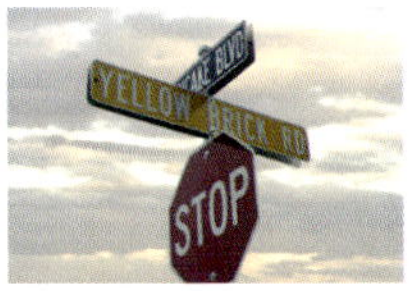
Eva Worobiec
Mamiya 645
80 mm
f16, 1/60th
Kodak Portra 400

40 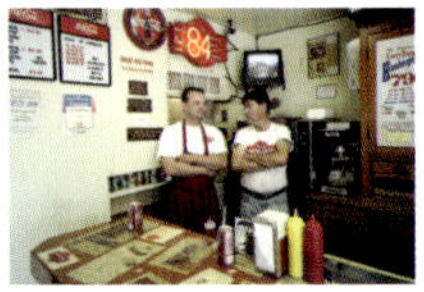
Tony Worobiec
Mamiya 7
43 mm
f8, 1/60th, flash
Kodak Portra 400

41 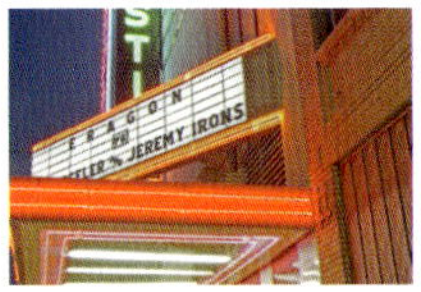
Tony Worobiec
Mamiya 7
43 mm
f11, 2 min
Fuji Reala 100

42 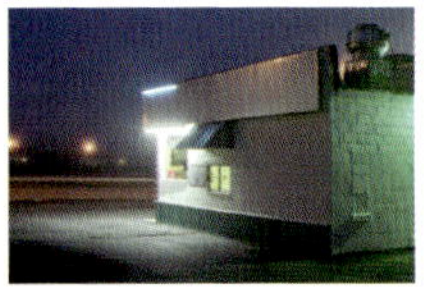
Tony Worobiec
Pentax 67
90 mm
f22, 4 min
Fuji Reala 100

43
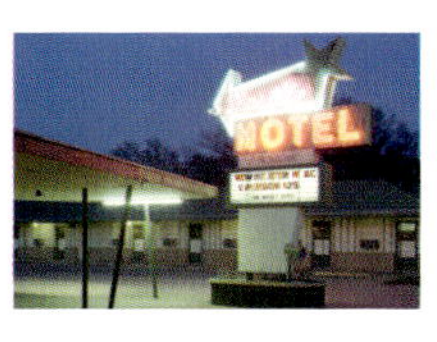
Eva Worobiec
Mamiya 645
45 mm
f16, 15 sec
Fuji Pro 160

44
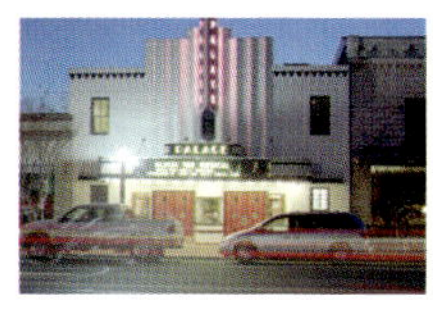
Tony Worobiec
Pentax 67
90 mm
f16, 2 min
Fuji Reala 100

45
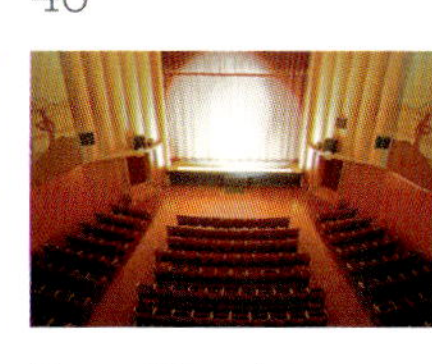
Tony Worobiec
Pentax 67
90 mm
f16, 2 min
Fuji Reala 100

46
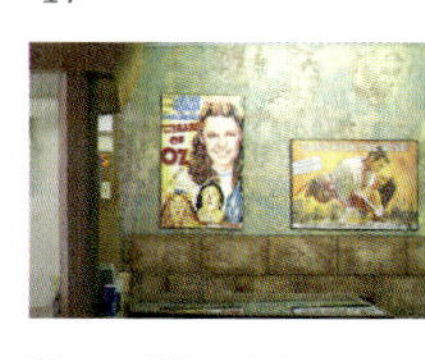
Tony Worobiec
Mamiya 7
43 mm
f16, 2 min
Fuji Pro 160

47
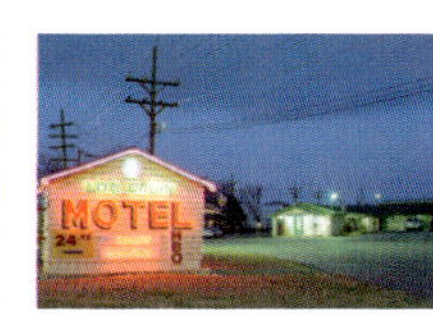
Tony Worobiec
Pentax 67
90 mm
f11, 4 sec
Fuji Reala 100

48
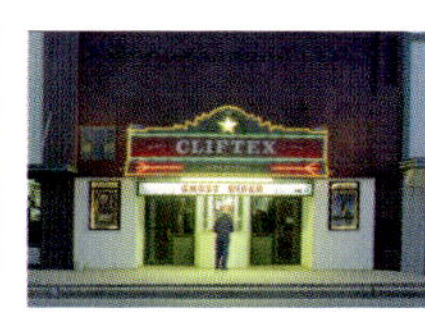
Tony Worobiec
Pentax 67
90 mm
f16, 1 min
Fuji Reala 100

49
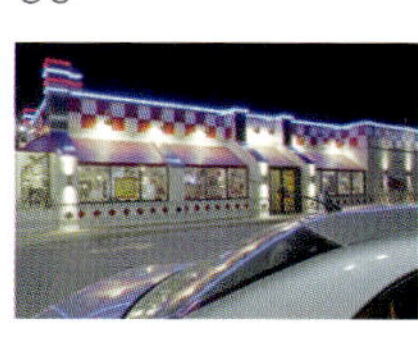
Tony Worobiec
Pentax 67
200 mm
f32, 2 sec
Fuji Reala 100

50
Eva Worobiec
Fuji 69
65 mm
f22, 2 min
Fuji Reala 100

51
Tony Worobiec
Pentax 67
45 mm
f16, 2 sec
Fuji Reala 100

52
Eva Worobiec
Fuji 69
65 mm
f16, 3 min
Fuji Reala 100

53
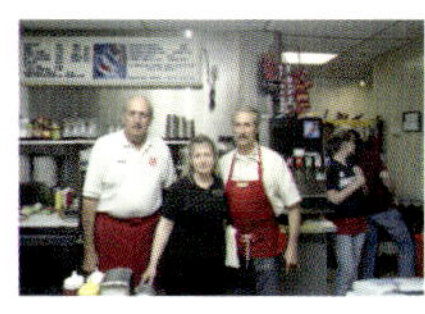
Eva Worobiec
Fuji 69
65 mm
f16, 2 min
Fuji Reala 100

54
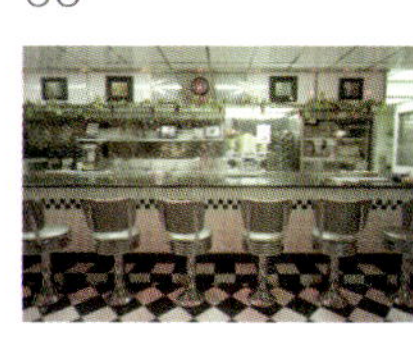
Tony Worobiec
Mamiya 7
80 mm
f11, 1/60th, flash
Fuji Pro 160

55
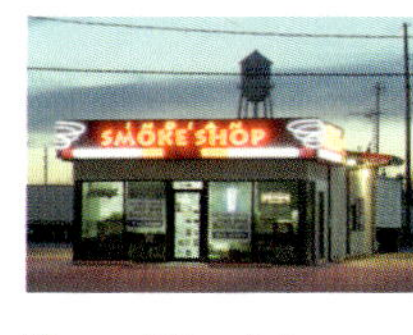
Tony Worobiec
Pentax 67
45 mm
f16, 1 sec
Fuji Reala 100

56
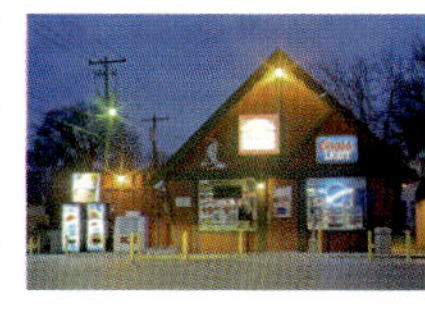
Tony Worobiec
Pentax 67
200 mm
f32, 2 min
Fuji Reala 100

57
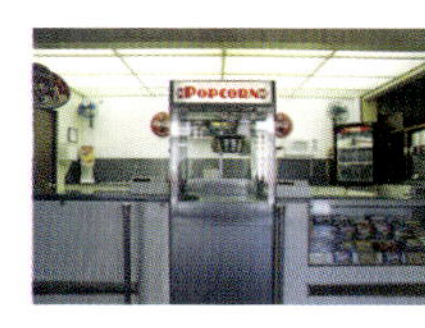
Eva Worobiec
Mamiya 645
80 mm
f16, 1 min
Fuji Reala 100

58
Tony Worobiec
Pentax 67
90 mm
f16, 2 sec
Fuji Reala 100

59
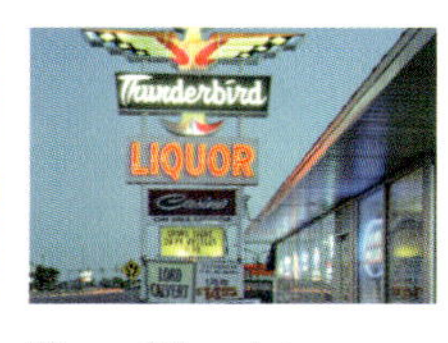
Tony Worobiec
Pentax 67
200 mm
f32, 2 min
Fuji Reala 100

62
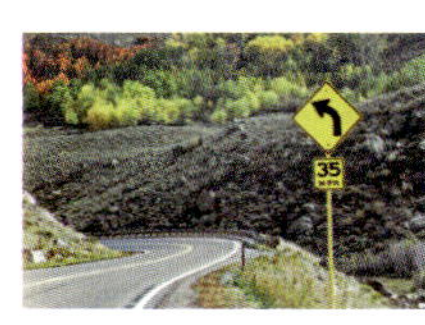
Tony Worobiec
Mamiya 7
80 mm
f16, 3 min
Fuji Reala 100

63
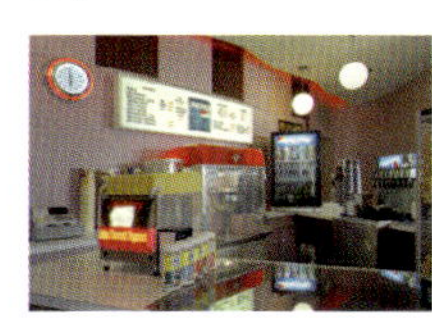
Tony Worobiec
Mamiya C330
250 mm
f45, 0.5 sec
Fuji Reala 100

64
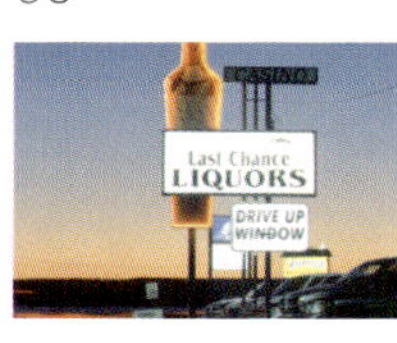
Eva Worobiec
Fuji 69
65 mm
f22, 4 sec
Fuji Reala 100

65
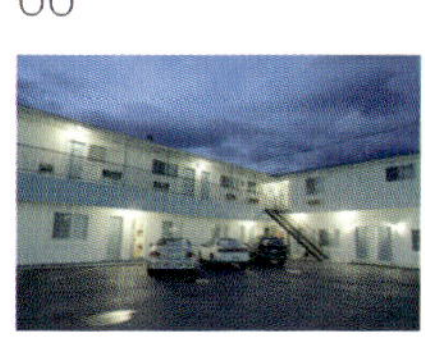
Tony Worobiec
Pentax 67
200 mm
f45, 4 min
Fuji Reala 100

66
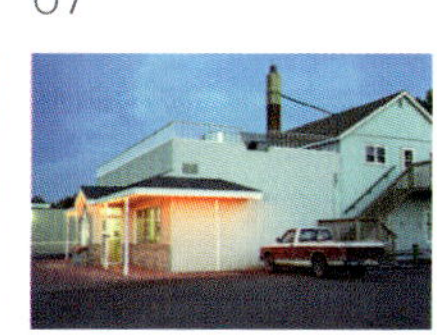
Tony Worobiec
Mamiya 7
43 mm
f16, 3 min
Fuji Reala 100

67
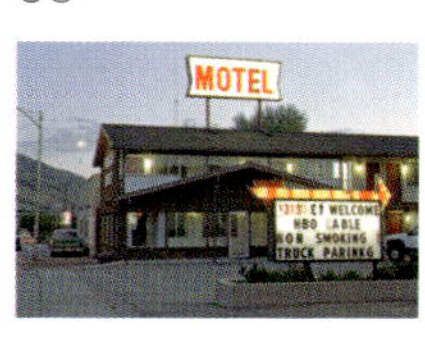
Tony Worobiec
Mamiya 7
80 mm
f16, 30 sec
Fuji Reala 100

68
Tony Worobiec
Mamiya C330
80 mm
f16, 90 sec
Fuji Reala 100

69
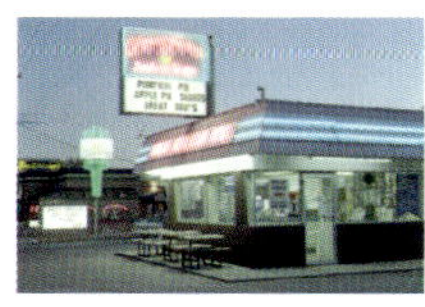
Tony Worobiec
Pentax 67
45 mm
f16, 4 min
Fuji Reala 100

70
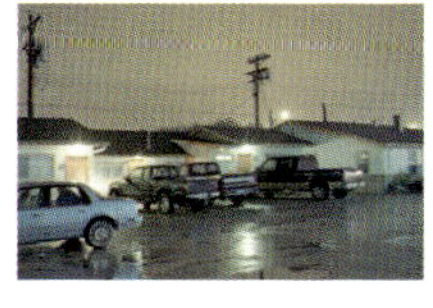
Tony Worobiec
Mamiya C330
80 mm
f16, 90 sec
Fuji Reala 100

71
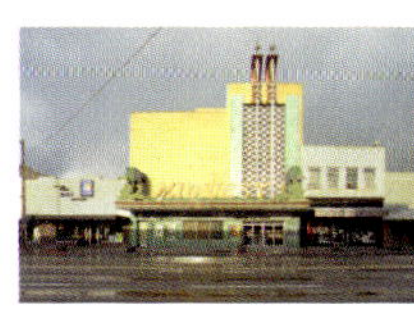
Tony Worobiec
Pentax 67
80 mm
f16, 5 min
Fuji Pro 160

72
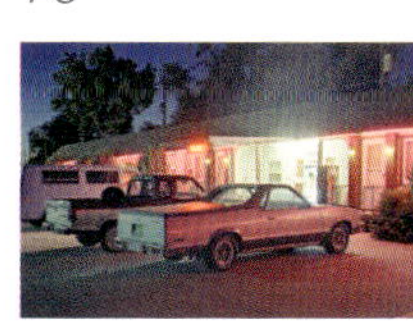
Eva Worobiec
Mamiya 645
45 mm
f22, 1 sec
Kodak Portra 160

73
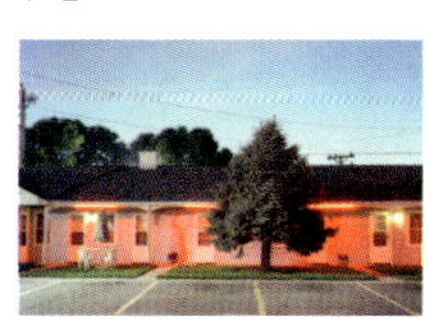
Tony Worobiec
Mamiya 7
80 mm
f16, 3 min
Fuji Pro 160

74
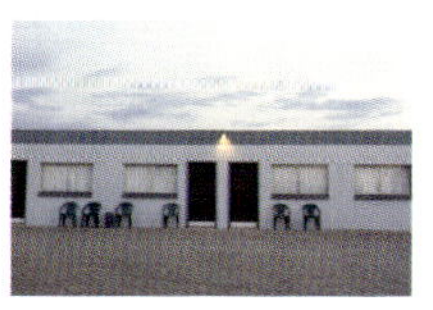
Tony Worobiec
Pentax 67
80 mm
f16, 30 sec
Fuji Reala 100

75
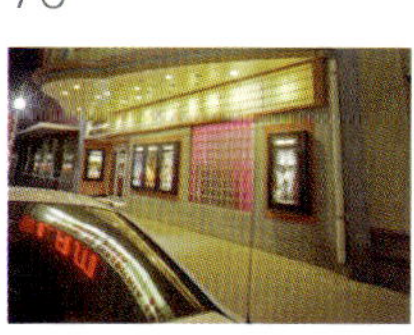
Tony Worobiec
Pentax 67
200 mm
f22, 1 sec
Fuji Reala 100

76
Tony Worobiec
Mamiya 7
43 mm
f16, 3 min
Fuji Pro 160

77
Eva Worobiec
Fuji 69
65 mm
f16, 2 min
Fuji Reala 100

78
Eva Worobiec
Fuji 69
65 mm
f16, 1 min
Fuji Pro 160

79
Tony Worobiec
Mamiya 7
80 mm
f16, 30 sec
Fuji Reala 100

80
Tony Worobiec
Pentax 67
45 mm
f16, 3 min
Fuji Reala 100

81
Tony Worobiec
Pentax 67
90 mm
f16, 8 sec
Fuji Pro 160

84
Eva Worobiec
Mamiya 645
50 mm
f16, 1 min
Kodak Portra 400

85
Tony Worobiec
Mamiya 7
80 mm
f16, 2 min
Fuji Reala 100

86
Tony Worobiec
Pentax 67
90 mm
f22, 1/125th
Kodak Portra 400

87
Tony Worobiec
Mamiya 7
43 mm
f16, 15 sec
Fuji Reala 100

88
Tony Worobiec
Pentax 67
90 mm
f11, 1/125th, flash
Kodak Portra 400

89
Tony Worobiec
Mamiya 7
80 mm
f16, 2 min
Fuji Reala 100

90
Eva Worobiec
Fuji 69
65 mm
f16, 2 min
Fuji Reala 100

91
Tony Worobiec
Mamiya 7
80 mm
f22, 60 sec
Fuji Reala 100

92
Eva Worobiec
Mamiya 645
50 mm
f16, 4 sec
NPH 400

93
Tony Worobiec
Pentax 67
90 mm
f16, 2 sec
Fuji Pro 160

94
Tony Worobiec
Mamiya 7
43 mm
f16, 3 min
Fuji Reala 100

95
Tony Worobiec
Pentax 67
45 mm
f16, 2 min
Fuji Reala 100

96
Eva Worobiec
Fuji 69
65 mm
f22, 1 min
Fuji Pro 160

97
Eva Worobiec
Mamiya 645
80 mm
f22, 1 min
NPH 400

98
Eva Worobiec
Fuji 69
65 mm
f16, 3 min
Fuji Reala 100

99
Tony Worobiec
Pentax 67
90 mm
f11, 30 sec
Fuji Pro 160

100
Tony Worobiec
Pentax 67
90 mm
f16, 2 min
Fuji Pro 160

101
Tony Worobiec
Pentax 67
90 mm
f11, 2 sec
Fuji Pro 160

102
Tony Worobiec
Mamiya 7
80 mm
f22, 4 min
Fuji Reala 100

103
Eva Worobiec
Mamiya 645
80 mm
f16, 8 sec
Kodak Portra 400

104
Tony Worobiec
Pentax 67
90 mm
f22, 30 sec
Fuji Pro 160

105
Eva Worobiec
Fuji 69
65 mm
f16, 1 sec
Fuji Reala 100

108
Tony Worobiec
Pentax 67
90 mm
f16, 1/125th
Fuji Pro 160

109
Tony Worobiec
Mamiya 7
80 mm
f16, 1/125
Fuji Reala 100

110
Tony Worobiec
Pentax 67
200 mm
f32, 15 sec
Fuji Pro 160

111
Eva Worobiec
Mamiya 645
50 mm
f16, 30 sec
Kodak Portra 400

112
Eva Worobiec
Mamiya 645
45 mm
f16, 1/30th
Fuji Reala 100

113
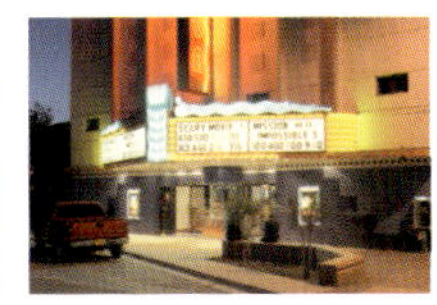
Eva Worobiec
Mamiya 645
80 mm
f16, 1/15th
Kodak Portra 400

114/115
Tony Worobiec
Mamiya 7
80 mm
f22, 4 sec, light
Kodak Portra 400

116
Eva Worobiec
Mamiya 645
45 mm
f16, 1 min
Kodak Portra 400

117
Eva Worobiec
Mamiya 645
50 mm
f16, 1 min
Fuji Reala 100

118
Tony Worobiec
Mamiya 7
43 mm
f16, 1 min
Fuji Reala 100

119
Tony Worobiec
Pentax 67
45 mm
f16, 2 min
Fuji Pro 160

120
Tony Worobiec
Pentax 67
200 mm
f32, 1.5 min
Fuji Reala 100

121
Eva Worobiec
Fuji 69
65 mm
f16, 15 sec
Fuji Pro 160

122
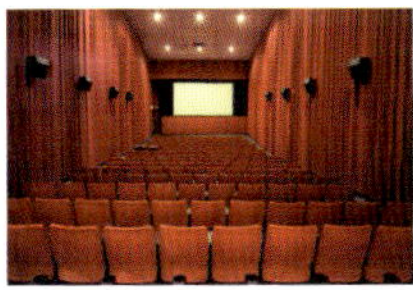
Tony Worobiec
Mamiya 7
43 mm
f22, 3min
Fuji Reala 100

123
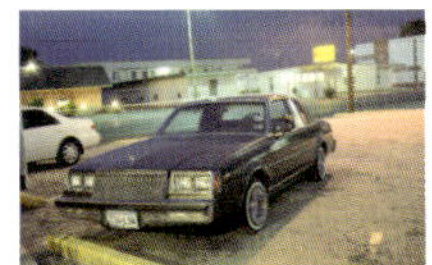
Tony Worobiec
Mamiya 7
43 mm
f16, 30 sec
Fuji Pro 160

pagenumber

photographer
camera
lens
aperture, exp. time
film

Technical Details

Tony Worobiec

Virtually all my images were taken on either a Pentax 67 or a Mamiya 7 camera. In the case of the Pentax, I would most commonly use either a 90mm or a 200mm lens for most of the exterior shots, whilst many of the interiors were taken with a 45mm lens. When using the Mamiya, I generally opted for either an 80 mm or a 43 mm lens. As long exposures were often required, it was more appropriate to use colour negative film, particularly Fuji Reala 100, which handles the problems of colour shift extremely well. I avoided the temptation of exposing tungsten balanced film as I prefer the almost "surreal" colours a daylight type film produces in these unusual lighting conditions. As the overwhelming majority of the images required long exposures (indeed many of the shots were in excess of three minutes), most were taken using a Gitzo G 212 tripod.

Eva Worobiec

With its interchangeable backs and wide variety of lenses, I opted for a Mamiya 645 system, although where a more "panoramic" approach was required I would also use a Fuji GSW690III camera. Using this larger format also added to the quality of some of the images. With the Mamiya the most suitable lenses were a 45mm wide-angle and the 80 mm standard; where necessary I took several images on a 50mm shift lens in order to correct converging verticals.

Whereas this could be done in a fairly leisurely manner for interior shots, the quickly-changing evening light would often preclude too much lens changing. Obviously, I was reliant on my trusty Manfrotto throughout, and could not have done without my versatile Gitzo ball-and-socket head either. Having access to two film backs, I was able to select fast or slow films according to the varying scenarios and found that Kodak Portra 400, Portra 160 and Fujicolor Pro 160 coped with the demands, as well as the ever dependable Fuji Reala 100.

Reciprocity was a constant worry for both of us and in particular the danger of under-exposure. We overcame this by taking frequent readings using hand-held light meters and then doubling the exposure. Initially we considered bracketing, but because of the lengthy exposures required and the rapidly diminishing light, this was a luxury we were rarely able to consider.

It was our aim to take most of our photographs at dusk or dawn in order to illustrate the neon lighting in its full resplendence. To achieve a tonal balance between the dimming sky and the emerging neon, we rarely had more than a 20 minute "window of opportunity". Also, to ensure that we were in the right place at the right time, we would regularly visit potential sites many hours beforehand, trying to assess which would offer the best opportunities. As the majority of the exposures were extremely lengthy, incidental figures moving across the frame during an exposure rarely caused problems as they were often too quick to register on film. Moving vehicles and their trailing lights, however, posed a different challenge: sometimes we were content to leave them in, but when we did wish to exclude them, simply covering the lens with a dark card as they passed by proved to be most effective.

Acknowledgements

We are often asked both by fellow Brits and Americans why we keep returning to the United States to pursue our photographic interests. With this particular project, which grew out of our earlier "Ghosts in the Wilderness" book, we were obviously drawn to capture those elements of a popular culture which have imbued our own lives, albeit on the other side of the "pond". Just as significant, however, have been the immense enthusiasm, kindness and hospitality shown to us by everyone we have encountered on our travels, whether they were the owners or staff in motels, diners and cinemas, or people we simply encountered in the street who pointed us in the right direction.

There is not space here to thank everyone personally, but we would like to mention in particular the following: Ed Alcorn, Sonja Asendorf, Kammie Cathcart, Richard Dudley, Maria Escamiela, John Eberly, Dodie Evans, Cecil George, Vicky Hayes-Walworth, Bobbie-Jo Heck, Betty Hukill, Ronny Jones, John Joy, Bill Kinder, Ramon Martinez, Roy McDowell, Carol Millirons, Gene Oliver, Mr and Mrs Nayan Patel, Cole Reed, Ron Richardson, Rick and Tina Ross, Barbara Smith and Alison and W. Leon Smith.

We are also grateful to the staff at Monarch Photo, Fargo, the Armadillo Camera Shop in Lubbock and Moler's Camera, Wichita, for keeping us stocked with 120 roll film and helping us to have crucial films processed in double-quick time. As ever, we thank Andrew Potter and his team at Ashley Colour Labs, Poole, and Cameron Brown for his support in the production of this book. Finally, we must express our appreciation to a good friend, Peter Baker, for his advice on the text, and to Stefan Nekuda for the quality of the design and his seemingly inexhaustible patience. We thank you all.